Current Topics in Microbiology and Immunology

119

Editors

Images of Biologically Active Structures in the Immune System

Their Use in Biology and Medicine

Edited by
H. Koprowski and F. Melchers

With 22 Figures

Springer-Verlag
Berlin Heidelberg NewYork Tokyo

Professor Dr. HILARY KOPROWSKI
Director
The Wistar Institute
36th Street at Spruce
Philadelphia, PA 19104, USA

Professor Dr. FRITZ MELCHERS
Basel Institute for Immunology
Grenzacherstraße 487
CH-4058 Basel

ISBN-13: 978-3-642-70677-6 e-ISBN-13: 978-3-642-70675-2
DOI: 10.1007/978-3-642-70675-2

Library of Congress Catalog Card Number 15-12910
Softcover reprint of the hardcover 1st edition 1985

Typesetting, printing and bookbinding:
Universitätsdruckerei H. Stürtz AG, Würzburg
2123/3130-543210

Preface

The diversity of antigen-binding structures of antibody molecules is so vast that every conceivable antigen can be bound by an antibody molecule within the immune system. This is true even for the antigen binding sites of antibodies called idiotypes, which are bound by complementary binding sites of other antibodies called anti-idiotypes. Thus, anti-idiotypes are structural homologues of antigens. These idiotypic-anti-idiotypic interactions constitute a network within the immune system. Since one lymphocyte produces only one type of antibody molecule, this network is in fact a network of cells. We expect that the network is functional: the appearance of antigen will disturb the equilibrium of the network at the point where it competes with the anti-idiotypic lymphocyte for binding to the idiotypic lymphocyte. It has been known for quite some time that anti-idiotypic antibody can be used to prime the immune system for memory to an antigen that it has never seen. This phenomenon is now being explored for possible use in immunization against viruses, bacteria, parasites and tumors as well as for the modulation of autoimmunity. The ability of anti-idiotypes to mimic, both antigenically and functionally, the corresponding biologically active molecules seen by an idiotypic antibody was first demonstrated for the hormone insulin and is now being observed in many other systems. The papers assembled in this volume bring the reader to the cutting edge of the potential practical applications of the network theory of the immune system. They relate the search for molecules and their biological functions and for the regulation of the immune response itself in autoimmunity and vaccination.

Summer 1985 HILARY KOPROWSKI, FRITZ MELCHERS

Table of Contents

Indexed in Current Contents

List of Contributors

Brait, M., Université Libre de Bruxelles, Laboratoire de Physiologie Animale, Département de Biologie Moléculaire, 67, rue des Chevaux, B-1640 Rhode-St.-Genèse

Bruyns, C., Université Libre de Bruxelles, Laboratoire de Physiologie Animale, Département de Biologie Moléculaire, 67, rue des Chevaux, B-1640 Rhode-St.-Genèse

Chamat, S., Laboratory of Molecular Immunology, CNRS, Institut Jacques Monod, University Paris VII, 2, place Jussieu, F-75251 Paris Cedex 05

DeFreitas, E., The Wistar Institute of Anatomy and Biology, 36th Street at Spruce, Philadelphia, PA 19104, USA

Demeur, C., Université Libre de Bruxelles, Laboratoire de Physiologie Animale, Département de Biologie Moléculaire, 67, rue des Chevaux, B-1640 Rhode-St. Genèse

Dubois, P., Université Libre de Bruxelles, Laboratoire de Physiologie Animale, Département de Biologie Moléculaire, 67, rue des Chevaux, B-1640 Rhode-St.-Genèse

Francotte, M., Université Libre de Bruxelles, Laboratoire de Physiologie Animale, Département de Biologie Moléculaire, 67, rue de Chevaux, B-1640 Rhode-St.-Genèse

Franssen, J.D., Université Libre de Bruxelles, Laboratoire de Physiologie Animale, Département de Biologie Moléculaire, 67, rue des Chevaux, B-1640 Rhode-St.-Genèse

Guillet, J.G., Laboratory of Molecular Immunology, CNRS, Institut Jacques Monod, University Paris VII, 2, place Jussieu, F-75251 Paris Cedex 05

Herlyn, D., The Wistar Institute of Anatomy and Biology, 36th Street at Spruce, Philadelphia, PA 19104, USA

Herlyn, M., The Wistar Institute of Anatomy and Biology, 36th Street at Spruce, Philadelphia, PA 19104, USA

Hiernaux, J., Université Libre de Bruxelles, Laboratoire de Physiologie Animale, Département de Biologie Moléculaire, 67, rue des Chevaux, B-1640 Rhode-St.-Genèse

Hoebeke, J., Laboratory of Molecular Immunology, CNRS, Institut Jacques Monod, University Paris VII, 2, place Jussieu, F-75251 Paris Cedex 05

Katz, D.H., The Department of Immunology, Medical

Biology Institute, 11077 North Torrey Pines Road, La Jolla, CA 92037, USA
KENNEDY, R.C., Virology and Immunology Department, Southwest Foundation for Biomedical Research, San Antonio, TX 78284, USA
KOPROWSKI, H., The Wistar Institute of Anatomy and Biology, 36th Street at Spruce, Philadelphia, PA 19104, USA
LEO, O., Université Libre de Bruxelles, Laboratoire de Physiologie Animale, Département de Biologie Moléculaire, 67, rue des Chevaux, B-1640 Rhode-St.-Genèse
LUBECK, M., The Wistar Institute of Anatomy and Biology, 36th Street at Spruce, Philadelphia, PA 19104, USA
MARVEL, J., Université Libre de Bruxelles, Laboratoire de Physiologie Animale, Département de Biologie Moléculaire, 67, rue des Chevaux, B-1640 Rhode-St.-Genèse
MEYERS, P., Université Libre de Bruxelles, Laboratoire de Physiologie Animale, Département de Biologie Moléculaire, 67, rue des Chevaux B-1640 Rhode-St.-Genèse
MOSER, M., Université Libre de Bruxelles, Laboratoire de Physiologie Animale, Département de Biologie Moléculaire, 67, rue de Chevaux, B-1640 Rhode St.-Genèse
OSTERHAUS, A.D.M.E., The National Institute of Public Health and Environmental Hygiene, P.O. Box 1, NL-3720 BA Bilthoven
REAGAN, K.J., The Wistar Institute of Anatomy and Biology, 36th Street at Spruce, Philadelphia, PA 19104, USA
SACKS, D.L., Laboratory of Parasitic Diseases, National Institute of Allergy and Infectious Diseases, National Institutes of Health, Bethesda, MD 20205, USA
SEARS, H., The Wistar Institute of Anatomy and Biology, 36th Street at Spruce, Philadelphia, PA 19104, USA
SLAOUI, M., Université Libre de Bruxelles, Laboratoire de Physiologie Animale, Département de Biologie Moléculaire, 67, rue des Chevaux, B-1640 Rhode-St.-Genèse
STEIN, K.E., Division of Bacterial Products, Office of Biologics Research and Review, Center for Drugs and Biologics, Food and Drug Administration, 8800 Rockville Pike, Bethesda, MD 20205, USA
STROSBERG, A.D., Laboratory of Molecular Immunology, CNRS, Institut Jacques Monod, University Paris VII, 2, place Jussieu, F-75251 Paris Cedex 05
SUZUKI, H., The Wistar Institute of Anatomy and Biology, 36th Street at Spruce, Philadelphia, PA 19104, USA
TASSIGNON, J., Université Libre de Bruxelles, Laboratoire de Physiologie Animale, Département de Biologie Moléculaire, 67, rue des Chevaux, B-1640 Rhode-St.-Genèse
URBAIN, J., Université Libre de Bruxelles, Laboratoire de

Physiologie Animale, Département de Biologie, Moléculaire, 67, rue des Chevaux, B-1640 Rhode-St.-Genèse

URBAIN-VANSANTEN, G., Université Libre de Bruxelles, Laboratoire de Physiologie Animale, Département de Biologie Moléculaire, 67, rue des Chevaux, B-1640 Rhode-St.-Genèse

UYTDEHAAG, F.G.C.M., The National Institute of Public Health and Environmental Hygiene, P.O. Box 1, NL-3720 BA Bilthoven

VAN ACKER, A., Université Libre de Bruxelles, Laboratoire de Physiologie Animale, Département de Biologie Moléculaire, 67, rue des Chevaux, B-1640 Rhode-St.-Genèse

Wikler, M., Université Libre de Bruxelles, Laboratoire de Physiologie Animale, Département de Biologie Moléculaire, 67, rue des Chevaux, B-1640 Rhode-St.-Genèse

WILLEMS, F., Université Libre de Bruxelles, Laboratoire de Physiologie Animale, Département de Biologie Moléculaire, 67, rue des Chevaux, B-1640 Rhode-St.-Genèse

WUILMART, C., Université Libre de Bruxelles, Laboratoire de Physiologie Animale, Département de Biologie Moléculaire, 67, rue des Chevaux, B-1640 Rhode-St.-Genese

ZANETTI, M., The Department of Immunology, Medical Biology Institute, 11077 North Torrey Pines Road, La Jolla, CA 92037, USA

Idiotype Networks in Hepatitis B Virus Infections*

R.C. KENNEDY

1 Introduction

The term idiotype (Id) or idiotypic determinant was originally proposed by OUDIN and MICHEL (1963) to designate antigenic determinants unique to a small set of antibody molecules. An Id defines the variable (V) region of the antibody molecule and often serves as a V region phenotypic marker. Because the V region of the antibody molecule also contains the antigen-binding region, the area of the antibody that makes contact with the antigen has been referred to as the paratope. It becomes important to remember that even within a single antibody molecule, different areas within the V region are capable of combining with different antigens. Thus, antigen binding to antibody requires sufficient complementarity between the antigen and antibody molecule to generate the attractive forces necessary for this reaction to occur. This complementarity or fitting of structural conformations plays a dual role in both antigen binding to an antibody molecule and antigen mimicry by potential anti-antibodies that recognize the conformation induced by the V region of an antibody molecule.

In early studies, antisera were generated against homogenous immunoglobulins, such as myeloma or Bence Jones proteins. After the appropriate adsorp-

* This work was supported in part by New Investigator Research Award A1-22307-01 from the National Institute of Allergy and Infectious Disease and in part by research contract DAMD 17-82-C 2155 from the US Army Medical Research and Development Command

Virology and Immunology Department, Southwest Foundation for Biomedical Research, San Antonio, TX 78284, USA

Current Topics in Microbiology and Immunology, Vol. 119

tions, these antisera reacted only with the myeloma protein used as the immunogen and not with other myelomas or conventional antibodies. Only rarely were exceptions to this antigenic individuality noted in these early studies. The first reports of antigenic individuality were based on a study of human myeloma proteins (LOHSS et al. 1953; SLATER et al. 1955) and extended to a series of conventional antibodies in humans (KUNKEL et al. 1963) and rabbits (OUDIN and MICHEL 1963). In each of these instances, the antisera generated against either the myeloma proteins or conventional antibodies recognized only the immunizing antibody. Cross-reactions with other myeloma proteins or antibodies were not detected. Thus, each antiserum detected unique or private determinants present on the V region of the immunogen. A term coined for the immunogenic potential of individual areas or determinants on the surface of the antigen-binding region that encompasses a single antibody molecule has been idiotope. Idiotope, by definition, implies monoclonality, and the unique collection of idiotopes on each antibody molecule is referred to as an Id. The Id is often defined by an antiidiotypic antibody (anti-Id or Ab2), whereby the Id can behave as an antigen and induce the production of antibodies against itself. It is noteworthy that the paratope may or may not be the same site as that recognized by the anti-Id. Therefore, the idiotope and paratope may represent distinct regions within the V region of an antibody molecule; the paratope being the area that binds antigen and the idiotope being the site that binds anti-Id.

The early studies on Id defined private or unique IdI determinants; however, shared Id that were common to more than one antibody preparation and that were found in numerous individuals have been termed by different investigators as public, shared, cross-reactive, or IdX determinants. Shared Id have been observed in a wide variety of species, with antibodies generated to numerous antigens. The detection of shared Id that are expressed on antibodies to a given antigen in different species (interspecies) suggested that these V region genes were conserved through evolution and most likely represent germ-line gene products. A distinction between shared and private Id is that shared Id appear to represent V region phenotypic markers and tend to be inherited, whereas private Id are not.

2 Anti-Id Modulation and Id Networks

The idea that the immune response to an antigen can be regulated via an Id-anti-Id network was first proposed by JERNE (1974). Id, located on or close to the antigen-binding site of both antibody molecules and lymphocyte antigen receptors, represents components of this network. Numerous studies have documented the manipulation of the immune response by injection of anti-Id. In particular, antigen exposure following injection of anti-Id resulted in either suppression of the Id-positive antigen-binding molecules (COSENZA and KOHLER 1972; EICHMAN 1974; HART et al. 1972) or increased Id expression and antigen-binding activity (BLUESTONE et al. 1981; CAZENAVE 1977; KELSOE et al. 1980; SACHS et al. 1981; TRENKNER and RIBLET 1975).

The generation of auto-anti-Id (reviewed by RODKEY 1980) provided further evidence that immune regulation involves a series of Id networks. Characterization of anti-anti-Id (Ab3) reagents revealed that in some instances these antisera were not able to bind the antigen used in the induction of the Id. However, within the framework of the network proposal, Ab3 can theoretically be produced with the specificity to bind to the original antigen by mimicking the Id. Thus, the anti-anti-Id has a structure or conformational fit that allows it to bind to the Id expressed on the surface of the anti-Id (Ab2). Alternatively, the Id of Ab1 itself has a structural conformation which allows the anti-Id favorable attractive forces for binding. Thus, the conformation of the anti-anti-Id (Ab3) should represent the mirror or internal image of the original Id of Ab1 and have the capacity to bind antigen. Several kinds of Ab3 molecules can be produced that differ in the ability to bind the original inducing antigen. Ab3 with antigen-binding capacity are induced by an Ab2 molecule, which has a structural conformation, such that it represents the internal image of the antigen.

In keeping with the concept that Id networks can regulate the immune response, T and B lymphocytes have been shown to share similar Id receptors (reviewed by BINZ and WIGZELL 1977 and KRAWINKEL et al. 1976). The generation of Id-specific suppressor T cells after administration of anti-Id in numerous murine antibody systems provided experimental evidence for cell regulation via Id networks (BONA and PAUL 1979; DOHI and NISONOFF 1979; EICHMANN 1975; YAMAMOTO et al. 1979). In addition, anti-Id generation of Id-specific helper T cells is also well documented (GLEASON et al. 1981; INADA et al. 1982; MILLER et al. 1981; SACHS et al. 1981; WOODLAND and CANTOR 1978). These data indicate that Id networks operate in regulation of the immune response to a given antigen.

3 Anti-Id Modulation of Viral Antigen Systems

The focus here will be on Id networks associated with hepatitis B virus infections, and these studies will be described in detail later in the chapter. However, much work has recently be done using anti-Id to modulate numerous viral antigen systems and a brief description of this work will be given below.

The first studies involving anti-Id for analysis of viral antigen systems were performed using tobacco mosaic virus (TMV) capsid protein (URBAIN et al. 1980). A single anti-Id, produced in a rabbit, against antibody to TMV detected a shared Id present on anti-TMV produced in rabbits, mice, horses, goats, and chickens immunized with TMV. In addition, this anti-Id induced the formation of anti-TMV when injected into mice without subsequent antigen challenge. The anti-Id represented the internal image of TMV and was capable of inducing an antibody response without ever injecting the antigen. Although TMV does not represent either a human or animal pathogen, these studies opened the door for others to study Id networks associated with viral pathogens. In addition, these studies implicated anti-Id as possible vaccines for viral infections via antigen mimicry.

Anti-Id have also been produced to characterize the hemagglutinin receptor for reovirus on both T lymphocytes and susceptible host cells (ERTL et al. 1982; NEPOM et al. 1982). Subsequent work indicated that anti-Id preparations could be used to block reovirus infection by competitively binding to cellular receptors (KAUFFMAN et al. 1983; NOSEWORTHY et al. 1983). The vaccine potential of a monoclonal anti-Id was demonstrated by its ability to induce cellular immunity to reovirus in BALB/c mice (SHARPE et al. 1984).

Anti-Id has also been demonstrated to induce neutralizing antibodies to rabies virus glycoproteins without subsequent antigen challenge (REAGAN et al. 1983). These studies again suggest the potential for an anti-Id based vaccine utilizing an anti-Id that may represent the internal image of a rabies virus glycoprotein important in inducing protective immunity.

Recently, a mouse monoclonal anti-Id was generated against a Sendai virus-specific helper T cell clone (ERTL and FINBERG 1984). This anti-Id was capable of inducing in mice an anti-Sendai virus immune response in vivo and further indicated that a T cell-defined anti-Id was capable of stimulating both B and T cell immunity which was induced by the anti-Id appeared to be less genetically restricted to the murine major histocompatability complex than was Sendai virus antigen-specific T cell immunity. More importantly, injection of anti-Id without subsequent antigen exposure produced immunity that was capable of protecting mice against a lethal infection with Sendai virus.

To date, the only viral antigen system where anti-Id has produced deleterious effects to the immune system has been with herpes simplex virus (HSV). The injection of anti-Id prior to a challenge with a 50% lethal dose of HSV resulted in a decrease in the survival time of mice (KENNEDY et al. 1984a). It was concluded from this study that the immune response to HSV in mice can be modulated through Id-anti-Id networks, which result in an increase in the pathogenicity of HSV infections. Although this study does implicate potential deleterious effects of anti-Id in modulating the immune response to infectious agents, numerous other studies (see above) have indicated the potential of anti-Id, that mimics viral antigens, as virus-free based vaccines.

4 Hepatitis B Virus

Hepatitis B virus (HBV) infects 175 million people yearly and represents a major worldwide health problem. The presence of HBV is often associated with the development of chronic liver disease and primary hepatocellular carcinoma (MAUPAS and MELNICK 1981). Fortunately, the majority of infected individuals recover completely, with the virus being eliminated and hepatic injury resolved. However, a chronic carrier state may occur in 5%–10% of HBV infections, whereby infectious HBV persists. In most patients who develop a chronic carrier disease, biochemical and histologic abnormalities are minimal to absent, but in others, hepatocellular injury may progress and lead to cirrhosis and carcinoma. It has been demonstrated that antibodies against hepatitis B surface antigen (anti-HBs), the surface or envelope material of HBV, are protective against infections, whereas antibody to the core antigen (HBcAg) has no protective effect. Serologically, hepatitis B surface antigen (HBsAg) contains within its structure a group-specific determinant(s) termed *a* and two sets of mutually exclusive allelic subtype determinants, *d* or *y* or *w* or *r*. Combinations of the *a* determinant with the various subtype determinants result in four major serotypes associated with HBsAg: *adw*, *ayw*, *adr*, and *ayr*. The group-specific *a* determinant(s) has been shown to induce protective antibodies against HBV (SZMUNESS et al. 1980).

At the present time, a vaccine for HBV is available and produced from the plasma of persons chronically infected with HBV by purifying HBsAg and treating the particles with disinfecting agents. Although this vaccine, licensed by Merck, Sharpe, and Dohme, has been proven safe and effective, its high cost and potentially limited availability preclude its use in developing countries, where HBV constitutes a major health problem. Thus, alternative approaches for an HBV vaccine are actively being pursued. The two major approaches under investigation for the preparation of a well-defined HBsAg-specific vaccine utilize recombinant DNA or synthetic peptide technology. A third possibility for an HBV vaccine lies within the confines of an anti-Id preparation that represents the internal image of HBsAg. Our laboratory has been actively pursuing the latter possibility.

5 Id Networks in HBV Infection: Characteristics of a Common Human Anti-HBs Id

Our impetus to study Id associated with antibodies to HBsAg (anti-HBs) came from the knowledge that anti-HBs, but not antibodies to the nucleocapsid core antigen, was protective against hepatitis B virus (HBV) infection. We initially generated four rabbit anti-Id antisera against affinity-purified anti-HBs from two different, naturally infected individuals. Each of the four anti-Id antisera detected a common anti-HBs Id (KENNEDY and DREESMAN 1983). A single anti-Id preparation was selected for further study. The common human Id was detected in purified anti-HBs from three individuals and also in anti-HBs positive sera obtained from six hemophilic patients. More recently, we have detected this Id in anti-HBs sera from 30 of 32 individuals who were either naturally infected with HBV or immunized with the HBsAg vaccine (KENNEDY et al., unpublished results). It is not known whether the two individuals who were negative for the anti-HBs Id failed to express that particular Id or expressed it at levels too low to be detected by the Id-anti-Id reaction. We are in the process of obtaining larger amounts of serum from these two individuals in order to affinity-purify the anti-HBs and test these preparations for inhibition of the Id-anti-Id reaction.

The ability of both HBsAg and a virus-derived HBsAg native polypeptide to inhibit the Id-anti-Id reaction suggested that the anti-HBs Id was associated with or was near the antibody-combining site. Id was detected because the anti-Id antisera did not interact effectively with IgG preparations from (a) the Id donor after removal of anti-HBs, (b) a pool of human sera negative for anti-HBs, and (c) an individual with a high level of antibody to herpes simplex virus. Attesting further to the Id specificity was the inability of the anti-Id to bind either HBsAg or the native HBsAg-derived polypeptide. These data indicated that we were detecting a common human anti-HBs Id associated with the antibody-combining site (KENNEDY and DREESMAN 1983).

Further characterization of the common Id revealed that it was induced by the group *a* determinant, because three HBsAg preparations purified from three pools of human plasma positive for HBsAg *adw*, *ayw*, or *adr* subtypes inhibited the Id-anti-Id reaction equally on a weight basis. We also tested the ability of HBsAg-derived polypeptides to inhibit the Id-anti-Id reaction and found that sodium dodecyl sulfate-denatured HBsAg viral polypeptides virtually lost their capacity to inhibit compared with a native polypeptide preparation. In addition, reduction of the disulfide bonds and alkylation of free thiol groups destroyed the ability of the native HBsAg-derived polypeptide to inhibit the Id-anti-Id reaction. These data suggested that the common anti-HBs Id was directed against a conformation-dependent group-specific *a* epitope (KENNEDY et al. 1982). Again, it is of note that the *a* determinant(s) of HBsAg induces protective immunity against HBV infection.

We have tested the ability of a cyclic synthetic peptide (peptide 1, containing amino acid residues 122–137 and being analogous to the major polypeptide, P25, of HBsAg) to inhibit the common Id-anti-Id reaction. On a molar basis, this peptide was 10^3-fold less efficient than intact HBsAg in inhibiting the Id-anti-Id reaction. The inability of peptide 1 to compete equally on a molar basis with HBsAg indicates that this peptide does not represent the complete *a* determinant that is recognized by the Id-anti-Id reaction and moreover suggests that other amino acid sequences are also important in defining the complete *a* epitope. However, inhibition of the Id-anti-Id reaction by peptide 1 suggested that this sequence is related to antigenic determinants responsible for eliciting a population of human anti-HBs expressing a common Id. The importance of conformation of the epitope was again tested by reducing the disulfide bond in peptide 1 and alkylating the free thiol groups. This treatment destroyed the ability of peptide 1 to inhibit the Id-anti-Id reaction (KENNEDY et al. 1983b). It could be argued that inhibition of the Id-anti-Id reaction with relatively large HBsAg particles was due to nonspecific steric hindrance that potentially could have resulted. However, the ability to inhibit the Id-anti-Id reaction with this relatively small molecule (mol. wt. 1704) strongly supports the above conclusion that the anti-Id recognized the antibody-combining site.

In addition, an interspecies Id cross-reaction was detected on anti-HBs produced in rabbits, guinea pigs, swine, goats, chimpanzees, and BALB/c mice immunized with HBsAg. Expression of the Id in sera from these other species was intrinsically associated with anti-HBs positive molecules. We also determined that anti-HBs from chickens successfully immunized with HBsAg failed to express the common Id (KENNEDY et al. 1983c). It was confirmed by adsorption studies that the interspecies Id was associated with anti-HBs. We also

Table 1. Expression of the interspecies Id by monoclonal anti-HBs that recognizes distinct determinants

Monoclonal antibody	Percentage inhibition	
	Cyclic peptide[a, b] 122–137	Id-anti-Id reaction[c]
A–2	36–88	24
A–4	31–64	25
A–7	0	0
A–12	53–67	26
A–16	0	0

[a] The cyclic peptide was utilized to inhibit the binding of the monoclonal anti-HBs preparations to intact HBsAg particles
[b] The inhibition values obtained using various inhibitor concentrations of the cyclic peptide, ranging from 500 ng to 10 μg, are shown
[c] The monoclonal anti-HBs preparation was used to inhibit the Id-anti-Id reaction

selected a series of mouse monoclonal anti-HBs preparations which differed in their ability to bind to cyclic peptide 1 and tested these anti-HBs preparations for the expression of the interspecies Id. Previous studies had suggested that the expression of the interspecies Id was determined by the specificity of a given monoclonal anti-HBs preparation for cyclic peptide 1 (IONESCU-MATIU et al. 1983). As shown in Table 1, the Id-anti-Id reaction appeared to recognize an anti-HBs response that was induced by determinants associated with cyclic peptide 1. Monoclonal anti-HBs preparations that either failed to bind cyclic peptide 1 or whose binding to intact HBsAg was not inhibited by cyclic peptide 1 failed to inhibit the Id-anti-Id reaction. Only those monoclonal anti-HBs preparations that were induced by determinants associated with cyclic peptide 1 expressed the interspecies Id. These findings indicate that the ability to respond to a distinct HBsAg epitope is highly conserved in mammalian species, but may not be shared by avian species. In addition, there appears to be a heterogeneity in the immune response to HBsAg, and the Id inhibitory capacity for the Id-anti-Id reaction does not appear to be based on total anti-HBs levels, but rather on the antibody response to the HBsAg epitopes associated with amino acid sequences 122–137. The Id-anti-Id reaction may represent a means by which the potential heterogeneity of the immune response to HBsAg can be measured.

6 Priming of the Anti-HBs Immune Response by Prior Injection of Anti-Id

The modulating effects of in vivo administration of anti-Id that recognized the common anti-HBs Id prior to antigenic challenge with HBsAg have been

Table 2. Priming of the anti-HBs response by prior injection of anti-Id[a]

First injection	Second injection	Mice (*n*)	Anti-HBs Response[b]	
			Range	Mean
Pre-IgG	Peptide 122–137	6	0–10	4
Anti-Id	Peptide 122–137	7	10–50	38
Pre-IgG	HBsAg	5	10–50	34
Anti-Id	HBsAg	10	1250–31250	9100
HBsAg	HBsAg	6	6250–156250	35620

[a] BALB/c mice received alum-precipitated antibodies (50 μg) or HBsAg (6 μg) on day 0, followed by either 6 μg HBsAg or 50 μg peptide on day 14; mice were bled on day 26

[b] Reciprocal of the anti-HBs endpoint titer, which has been described elsewhere (Kennedy and Dreesman 1984)

studied in mice at both the serum and cellular levels. The injection of anti-Id prior to HBsAg resulted in an increased number of spleen cells secreting IgM anti-HBs (Kennedy et al. 1983a). Spleen cells also were induced to secrete anti-HBs anti-Id only without antigen exposure. Serologically, it was determined that anti-Id given in saline induced predominantly an IgM anti-HBs response, whereas alum precipitated anti-Id produced an IgG anti-HBs response when administered prior to HBsAg (Kennedy and Dreesman 1984). Similarly, anti-Id were found to enhance the anti-HBs response to cyclic synthetic peptide 1 (Kennedy et al. 1984c). The injection of anti-Id prior to immunization with cyclic peptide 122–137 increased the anti-HBs response compared with groups of mice that received control antibodies prior to the peptide. Although the anti-Id in conjugation with the peptide produced low levels of anti-HBs compared with anti-Id and HBsAg, the anti-HBs titers in these mice were comparable with those obtained in mice receiving a single injection of HBsAg. It was noteworthy that a single injection of anti-Id failed to produce a detectable anti-HBs response and could only prime the antibody response to a subsequent challenge with antigen. Also, the anti-HBs response generated in mice receiving anti-Id and cyclic peptide 1 was associated with the group specific *a* determinant(s) of HBsAg. The data on anti-Id priming of the anti-HBs response are summarized in Table 2.

7 Anti-Id Bearing the Internal Image of HBsAg

In a previous study, we demonstrated that BALB/c mice receiving anti-Id alone generated IgG anti-HBs secreting spleen cells (Kennedy et al. 1983a). These data indicated that anti-Id alone could induce anti-HBs response. Mice that received two injections of anti-Id without subsequent HBsAg injection produced an anti-HBs response that was found to express the interspecies idiotype. In addition, the anti-Id induced anti-HBs recognized the group-specific *a* determinant(s) of HBsAg (Kennedy et al. 1984b). The injection of anti-Id alone produced an antibody response that appeared to serologically reflect the internal image of HBsAg. That is, the anti-HBs expressed the interspecies Id and recognized the group-specific *a* determinant. Alternatively, the potential role of anti-Id as vaccines for HBV infection was demonstrated by the fact that anti-Id immunization alone induced an anti-HBs response that shares an Id expressed in a human anti-HBs response which resulted from a natural HBV infection. This anti-HBs response protects these individuals from subsequent re-infection with HBV. Also, the specificity of the anti-HBs response produced by anti-Id injection was directed against a determinant(s) on HBsAg that is responsible for inducing protective immunity. Whether or not the anti-Id represents a possible vaccine for HBV must await further testing, since only chimpanzees and humans can be infected with human HBV. We are in the process of testing this possibility by anti-Id induction of anti-HBs in chimpanzees.

8 Does the Anti-Id Induced Anti-HBs Represent an Anti-Anti-Id Response?

The fact that the anti-Id preparation appears to represent the internal image of HBsAg raised the question of whether the anti-HBs response produced by anti-Id injection represented the induction of anti-anti-Id (Ab3). In order to answer this question, we selected two affinity-purified anti-Id preparations produced in different rabbits which appeared to represent the internal image of HBsAg by their capacity to induce anti-HBs in mice. The two rabbits used in the anti-Id generation were rested for a period of 14 months following their last injection of human anti-HBs Id. These rabbits were then immunized with

Table 3. Induction of an anti-anti-Id response with HBsAg binding activity in rabbits[a]

Rabbit	Antiserum obtained after:	Anti-HBs titer[b]
#1	Pre-immune	<2[c]
	Primary injection	<2
	Secondary injection	10
	Tertiary injection	50
	Quaternary injection	250
#2	Pre-immune	<2
	Primary injection	<2
	Secondary injection	50
	Tertiary injection	6250
	Quaternary injection	31250

[a] Each rabbit received 500 μg of its respective anti-Id preparation as an alum precipitate in complete Freund's adjuvant at monthly intervals. Serum was obtained prior to each immunization

[b] Reciprocal of endpoint anti-HBs titer using a solid-phase radioimmunoassay with HBsAg coated wells and ^{125}I-labeled goat antirabbit-γ-globulin. Each serum was diluted in fivefold increments using 10% normal goat serum as a dilutent

[c] No anti-HBs activity at a 1:2 dilution of serum

their respective affinity purified anti-Id preparations. Following four injections of 500 µg of their own anti-Id preparations, both rabbits produced a detectable anti-HBs response (Table 3). No anti-HBs activity was detectable prior to anti-Id immunization, although antibody to human IgG was still present. It was anticipated that these rabbits would share isotypic determinants with the anti-Id preparations which were produced in response to Id injection. Thus, the only response to the anti-Id injection should have been to the Id on the anti-Id or anti-anti-Id. If the anti-Id represents a true internal image due to complementarity, then differences in species should not affect the internal image of the anti-Id and the induction of an anti-anti-Id response in the rabbits should bind HBsAg. Although the specificity of the anti-anti-Id response other than HBsAg binding activity is not yet shown, the data strongly suggest that the anti-Id induced anti-HBs activity was the result of an Ab3 response.

9 Concluding Remarks

This chapter discusses and characterizes idiotype networks observed in hepatitis B virus infection or HBsAg immunization. In addition, the potential use of anti-Id that represents the internal image of HBsAg as possible vaccine or vaccine primers for HBV was also discussed. However, one should be aware of the problems associated with an anti-Id based vaccine. In certain instances, anti-Id induced protection was reported to be genetically restricted (Sacks and Sher 1983). Yet, in this particular study, the anti-Id did not appear to represent an internal image of protective protozoan epitopes. It is anticipated that if an anti-Id mimics an antigen via complementarity in its three-dimensional structure, then, unless the anti-Id is processed in vivo as an antigen via a different mechanism than the original antigen itself, the anti-Id should be genetically restricted only when the immune response to original antigen is restricted. Indeed, anti-Id induced immunity to Sendai virus appears to be less genetically restrictive than that induced by the virus (Ertl et al. 1984). Conversely, the injection of a heterologous antibody preparation (either rabbit or mouse) into humans has obvious disadvantages. Technological advances have been made whereby human hybridoma anti-Id vaccines could be made in quantities that might make the vaccine cost effective. However, the injection of homologous antibody preparations could potentially induce adverse allergic or immune complex reactions, especially when administered in multiple vaccinations over a period of time. There are also questions regarding how long lasting is anti-Id induced immunity. Where then does the future of anti-Id vaccines lie? The resolution of the problems associated with heterologous and homologous antibody immunogens may be found in the use of synthetic peptides with a predefined specificity and conformation (H. Kohler, personal communication). Once the three-dimensional structure of anti-Id bearing an internal image of an antigen is unraveled, synthetic peptides can be produced with a complementarity similar to that of the antigen. With regard to the problems associated with inducing long-term immunity, these are the same problems facing proponents of synthetic peptides as alternative vaccine candidates. Hopefully, this problem can be solved with the development of new synthetic adjuvants or understanding of the fine specificity of antigen presentation in T and B cell collaboration.

In conclusion, anti-Id appears to represent a viable alternative for vaccines where there is difficulty in obtaining adequate amounts of antigen. This would occur where the organism producing the infectious disease is difficult to obtain or grow in large numbers. Also anti-Id preparations are not infectious agents and may be useful when a conventionally attenuated vaccine has a high propensity for reverting to a virulent form. A final possibility where anti-Id may be advantageous as a vaccine is the instance where immunity to a single epitope on an infectious agent is adequate for inducing protection, especially if the infectious agent contains antigenic determinants in common with host tissue or body components. A vaccine directed to the entire infectious agent could potentially produce a deleterious auto-immune response against host tissues or other components, whereas the anti-Id would induce immunity against a single epitope.

Although there are still relatively few studies on the use of anti-Id as potential vaccines, anti-Id are being actively generated for the analysis of the immune response to numerous antigenic systems utilizing infectious agents that cause human and animal disease.

References

Binz H, Wigzell H (1977) Antigen-binding idiotypic T-lymphocyte receptors. Contemp Top Immunobiol 7:113–177

Bluestone JA, Sharrow SO, Epstein SL, Ozato K, Sachs DH (1981) Induction of anti-H-2 antibodies without alloantigen exposure by in vivo administration of anti-idiotype. Nature (Lond) 291:233–235

Bona C, Paul WE (1979) Cellular basis of regulation of expression of idiotype. I. T-suppressor cells specific for MOPC 460 idiotype regulate the expression of cells secreting anti-TNP antibodies bearing 460 idiotype. J Exp Med 149:592–600

Cazenave P-A (1977) Idiotypic-anti-idiotypic regulation of antibody synthesis in rabbits. Proc Natl Acad Sci USA 74:5122–5125

Cosenza H. Kohler H (1972) Specific suppression of the antibody response by antibodies to receptors. Proc Natl Acad Sci USA 69:2701–2705

Dohi Y, Nisonoff A (1979) Suppression of idiotype and generation of suppressor T cells with idiotype-conjugated thymocytes. J Exp Med 150:909–918

Eichmann K (1974) Idiotype suppression. I. Influence of the dose and of the effector functions of anti-idiotypic antibody on the production of an idiotype. Eur J Immunol 4:296–302

Eichmann K (1975) Idiotype suppression. II. Amplification of a suppressor T cell with anti-idiotypic activity. Eur J Immunol 5:511–517

Ertl HCJ, Finberg RW (1984) Sendai virus-specific T-cell clones: Induction of cytolytic T cells by an anti-idiotypic antibody directed against a helper T-cell clone. Proc Natl Acad Sci USA 81:2850–2854

Ertl HCJ, Greene MI, Noseworthy JH, Fields BN, Nepom JT, Spriggs DR, Finberg RW (1982) Identification of idiotypic receptors on reovirus-specific cytolytic T cells. Proc Natl Acad Sci USA 79:7479–7483

Ertl HCJ, Hornans E, Tournas S, Finberg RW (1984) Sendai virus-specific T cell clones. V Induction of a virus-specific response by antiidiotypic antibodies directed against a T helper cell clone. J Exp Med 159:1778–1783

Gleason K, Pierce S, Kohler H (1981) Generation of idiotype-specific T cell help through network perturbation. J Exp Med 153:924–935

Hart DA, Wang A-L, Pawlak LL, Nisonoff A (1972) Suppression of idiotypic specificities in adult mice by administration of antiidiotypic antibody. J Exp Med 135:1293–1300

Inada T, Yamamoto H, Bitoh S, Fujimoto S, Hamaoka T (1982) Detection and characterization of idiotype-specific enhancing cells generated in mice immunized with idiotype. J Immunol 129:2069–2073

Ionescu-Matiu I, Kennedy RC, Sparrow JT, Culwell AR, Sanchez Y, Melnick JL, Dreesman GR (1983) Epitopes associated with a synthetic hepatitis B surface antigen peptide. J Immunol 130:1947–1952

Jerne NK (1974) Towards a network theory of the immune system. Ann Immunol (Paris) 125c:373–389

Kauffman RS, Noseworthy JH, Nepom JT, Finberg R, Fields BN, Greene MI (1983) Cell receptors for the mammalian reovirus. II. Monoclonal anti-idiotypic antibody blocks viral binding to cells. J Immunol 131:2539–2541

Kelsoe G, Reth M, Rajewsky K (1980) Control of idiotype expression by monoclonal anti-idiotope antibodies. Immunol Rev 52:75–88

Kennedy RC, Dreesman GR (1983) Common idiotypic determinant associated with human antibodies to hepatitis B surface antigen. J Immunol 130:385–389

Kennedy RC, Dreesman GR (1984) Enhancement of the immune response to hepatitis B surface antigen: In vivo administration of antiidiotype induces anti-HBs that expresses a similar idiotype. J Exp Med 159:655–665

Kennedy RC, Sanchez Y, Ionescu-Matiu I, Melnick JL, Dreesman GR (1982) A common human anti-hepatitis B surface antigen idiotype is associated with the group *a* conformation-dependent antigenic determinant. Virology 122:219–221

Kennedy RC, Adler-Storthz K, Henkel RD, Sanchez Y, Melnick JL, Dreesman GR (1983a) Immune response to hepatitis B surface antigen: Enhancement by prior injection of antibodies to the idiotype. Science 221:853–855

Kennedy RC, Dreesman GR, Sparrow JT, Culwell AR, Sanchez Y, Ionescu-Matiu I, Hollinger FB, Melnick JL (1983b) Inhibition of a common human anti-hepatitis B surface antigen idiotype by a cyclic synthetic peptide. J Virol 46:653–655

Kennedy RC, Ionescu-Matiu I, Sanchez Y, Dreesman GR (1983c) Detection of interspecies idiotypic cross-reactions associated with antibodies to hepatitis B surface antigen. Eur J Immunol 13:232–235

Kennedy RC, Adler-Storthz K, Burns JW Sr, Henkel RD, Dreesman GR (1984a) Antiidiotype modulation of herpes simplex virus infection leading to increased pathogenicity. J Virol 50:951–953

Kennedy RC, Melnick JL, Dreesman GR (1984b) Antibody to hepatitis B virus induced by injecting antibodies to the idiotype. Science 223:930–931

Kennedy RC, Sparrow JT, Sanchez Y, Melnick JL, Dreesman GR (1984c) Enhancement of viral hepatitis B antibody (anti-HBs) response to a synthetic cyclic peptide by priming with anti-idiotype antibodies. Virology 136:247–252

Krawinkel U, Cramer M, Berek C, Hämmerling G, Black SJ, Rajewsky K, Eichmann K (1976) On the structure of the T-cell receptor for antigen. Cold Spring Harbor Symp Quant Biol 41:285–294

Kunkel HG, Mannik M, Williams RC (1963) Individual antigenic specificity of isolated antibodies. Science 140:1218–1219

Lohss F, Weiler E, Hillmann G (1953) Myolom-plasma-proteine, III. Mitteilung zur Immunochemie der γ-myelom-proteine. Z Naturforsch 8B:625–631

Maupas P, Melnick JL (eds) (1981) Hepatitis B virus and primary hepatocellular carcinoma. Prog Med Virol 27:1–210

Miller GG, Nadler PI, Asano Y, Hodes RJ, Sachs DH (1981) Induction of idiotype-bearing nuclease-specific helper T cells by in vivo treatment with anti-idiotype. J Exp Med 154:24–34

Nepom JT, Weiner HL, Dichter MA, Tardieu M, Spriggs DR, Gramm CF, Powers ML, Fields BN, Greene MI (1982) Identification of a hemagglutinin-specific idiotype associated with reovirus recognition shared by lymphoid and neural cells. J Exp Med 155:155–167

Noseworthy JH, Fields BN, Dichter MA, Sobotka C, Pizer E, Perry LL, Nepom JT, Greene MI

(1983) Cell receptors for mammalian reovirus. I. Syngeneic monoclonal anti-idiotypic antibody identifies a cell surface receptor for reovirus. J Immunol 131:2533–2538
Oudin J, Michel M (1963) Une nouvelle forme d'allotypie des globulines du sérum de lapin, apparemment liée à la fonction et à la spécificité anticorps. CR Séances Acad Sci 257:805–808
Reagan KJ, Wunner WH, Wiktor TJ, Koprowski H (1983) Anti-idiotypic antibodies induce neutralizing antibodies to rabies virus glycoprotein. J Virol 48:660–666
Rodkey LS (1980) Autoregulation of immune responses via idiotype network interactions. Microbiol Rev 44:631–659
Sachs DH, El-Gamil M, Miller G (1981) Genetic control of the immune response to staphylococcal nuclease. XI. Effects of *in vivo* administration of anti-idiotypic antibodies. Eur J Immunol 11:509–516
Sacks DL, Sher A (1983) Evidence that anti-idiotype induced immunity to experimental African trypanosomiasis is genetically restricted and requires recognition of combining site-related idiotopes. J Immunol 131:1511–1515
Sharpe AH, Gaulton GN, McDade KK, Fields BN, Greene MI (1984) Syngenic monoclonal antiidiotype can induce cellular immunity to reovirus. J Exp Med 160:1195–1205
Slater RJ, Ward SM, Kunkel HG (1955) Immunological relationships among myeloma proteins. J Exp Med 101:85–108
Szmuness W, Stevens CE, Harley EJ, Zang EA, Oleszko WR, William DC, Sadovsky R, Morrison JM, Kellner A (1980) Hepatitis B vaccine. Demonstration of efficacy in a controlled clinical trial in a high-risk population in the United States. N Engl J Med 303:833–841
Trenkner E, Riblet R (1975) Induction of antiphosphorylcholine antibody formation by anti-idiotypic antibodies. J Exp Med 142:1121–1132
Urbain J, Wikler M, Franssen JD, Collignon C (1977) Idiotypic regulation of the immune system by the induction of antibodies against anti-idiotypic antibodies. Proc Natl Acad Sci USA 74:5126–5130
Urbain J, Cazenave PA, Wikler M, Franssen JD, Mariame B, Leo O (1980) Idiotypic induction and immune networks. In: Fougereun M, Dausset J (eds) International Congress of Immunology, 4th, Paris, France, 1980. Immunology 80: progress in immunology IV. Academic, London, p 81
Woodland R, Cantor H (1978) Idiotype-specific T helper cells are required to induce idiotype-positive B memory cells to secrete antibody. Eur J Immunol 8:600–606
Yamamoto H, Nonaka M, Katz DH (1979) Suppression of hapten-specific delayed-type hypersensitivity responses in mice by idiotype-specific suppressor T cells after administration of anti-idiotypic antibodies. J Exp Med 150:818–829

Modulation of Immunity to Rabies Virus Induced by Anti-Idiotypic Antibodies

K.J. Reagan

1 Introduction

In the absence of effective postexposure therapies, viral diseases must be prevented by immunization with specific live-attenuated or killed vaccines. Nevertheless, complications from conventional vaccines, such as allergic responses to contaminating proteins, reversion of attenuated strains to virulence, or incomplete inactivation, must be recognized and the risks of vaccination must be balanced against the benefits. Vaccine design is now evolving to include only the minimal essential component of an infectious organism necessary to induce protective immunity. In recent years intense research has concentrated on recombinant DNA techniques and the chemical synthesis of short peptides of specific amino acid sequences as the ultimate in safe (genome-free) vaccines. However, their success will depend upon reproducing the immunogenicity and antigenicity of the original antigen. So far, synthetic peptides have not found general success in stimulating immunity to other than continuous antigenic determinants (which frequently are not the most biologically relevant) and are weakly immunogenic in the absence of adjuvants. Similarly, if the native conformation of an antigenic determinant epitope is formed by the interaction of more than one protein, recombinant technologies will also have difficulty. For these reasons, combined with the lack of detailed molecular information for many infectious organisms, there is incentive to explore alternative vaccination methods.

The Wistar Institute of Anatomy and Biology, 36th Street at Spruce, Philadelphia, PA 19104, USA
Present address: Glasgow Research Laboratory, E.I. DuPont de Nemours and Company, Inc., Wilmington, DE 19898, USA

Current Topics in Microbiology and Immunology, Vol. 119

This review details an approach to vaccination designed to induce immunity in animals with the minimal use or even absence of antigens of infectious origin, while stimulating specific antibodies to nonlinear epitopes. The procedure involves the use of anti-idiotypic antibodies (Ab2) directed against variable region (V) determinants in the heavy and/or light chains of other immunoglobulins (REAGAN et al. 1983, 1985).

The rabies virus provides an excellent model system for the development of new vaccines, particularly because we already have considerable knowledge of its molecular structure and pathogenesis. Among viral diseases, rabies is unique in that, due to its prolonged incubation period, vaccination postexposure is frequently successful. Current vaccination therapy has demonstrated that disease can be prevented by a combination of active and passive immunologic defenses, if provided before the virus is sequestered within the nervous system, and it is recognized that any effective vaccine for rabies must stimulate both the humoral and cellular arms of the immune system.

2 Background to the Use of Ab2 as a Vaccine

Classically, the immune system has been perceived to function by recognizing epitopes on cells or molecules that are foreign to an immunocompetent host, whereas immune responses to self-antigens were indicative of a disease state (auto-immunity) (RODKEY 1980). With the discovery of idiotypes (Id) (KUNKEL et al. 1963; OUDIN and MICHEL 1963) and the description of the Id network (JERNE 1974), this impression of the immune response has been modified to include the realization that self-reactivity is not necessarily detrimental but rather an integral part of immune regulation. JERNE (1974, 1975) suggests the existence of complementary sets of V domains which provide stimulatory or suppressive effects on lymphocytes bearing relevant Id receptors. Thus, within the immune system there is believed to be a steady state of mutual suppression between Id and anti-Id until an antigen arrives to upset the balance. The implication of this theory is that B cells can be triggered to synthesize antibodies in the absence of antigen exposure.

Sequential sets of antibodies involved in this Id network have been described by URBAIN et al. (1983). They include (a) Ab1, the Id present on an antibody in response to a given antigen; (b) Ab2, the anti-Id antibodies raised against Ab1; (c) Ab3, a diverse population of anti-anti-Id antibodies raised against Ab2. The Ab3 population may include (i) Ab3 that recognizes Id of Ab2 but nothing else, (ii) Ab3 that does not bind antigen but shares Id with Ab1, and (iii) Ab3 which shares an Id marker with Ab1 and also binds antigen. This latter subset contains the Ab1' which appears in large amounts after antigen injection in Ab3 animals. Finally, (d) Ab4 is anti-anti-anti-Id raised against Ab3 and has been shown to recognize determinants shared between Ab1 and Ab3.

Numerous investigators have documented the ability to manipulate immune responses by Ab2. In vivo administration of Ab2 followed by antigen exposure has often resulted in suppression of Id-bearing, antigen-binding molecules (COZENZA and KOHLER 1972; EICHMANN 1974; HART et al. 1972) perhaps due to the production of Id-specific T suppressor cells or by the deletion of helper T cells or precursor B cells bearing the relevant Id. However, enhanced Id expression and antigen binding has also been reported (see review, SACKS et al. 1983), possibly due to the stimulation of helper T cells (BOTTOMLY et al. 1978; TRENKNER and RIBLET 1975), the elimination of Id-specific suppressor T cells (BONA et al. 1978), or by antigen mimicry (KAUFFMAN et al. 1983). NISONOFF and LAMOYI (1981) first hypothesized the potential of using Ab2 as a vaccine for infectious diseases, yet direct evidence for a beneficial effect of Ab2 was not noted until SACKS et al. (1982) demonstrated the ability to immunize mice against *Tryponosoma rhodesiense* by prior treatment with allogenic Ab2 prepared against three protective monoclonal antibodies (mAb), each with a specificity for the major surface glycoprotein. The subsequent expansion of the data set with other infectious disease model systems (summarized in Table 1) has firmly established the potential of Ab2 modulation of specific immunity.

3 Ab2 as a Rabies Vaccine

Of the five structural proteins in the rabies virus, only the spike glycoprotein (G protein) is capable of inducing virus neutralizing (VN) antibody (Cox et al. 1977). In addition, G protein is involved in the generation of cell-mediated immunity (WIKTOR et al. 1974; MACFARLAN et al. 1984), neurovirulence (DIETZSCHOLD et al. 1983), attachment to cell surface receptors (WUNNER et al. 1984), and pH-dependent fusion of membranes during adsorptive endocytosis (REAGAN and WUNNER 1984). The isolated G protein has been shown to induce protective immunity (WIKTOR et al. 1973) and thus has been the target of efforts to produce a rabies-specific subunit vaccine (see review, WUNNER et al. 1983). Most significantly, a double-stranded cDNA copy of the ERA strain of rabies virus G protein mRNA recently was isolated, cloned, and sequenced (ANILIONIS et al. 1981). A full-length, glycosylated protein has been expressed upon insertion of this gene into a vaccinia virus vector system (KIENY et al. 1984). This protein reacts with a panel of anti-G protein mAb equally as well as G protein isolated from rabies virions does (KIENY et al. 1984). It induces high levels of VN antibody in animals, stimulates cytotoxic T cells and protects against severe intracerebral challenge with street strains of rabies virus (WIKTOR et al. 1984). However, although highly effective, the question about reintroduction of vaccinia virus to a world free from smallpox has not been resolved.

3.1 Preparation and Characterization of Ab2

In order to obtain functional Ab2, Id-bearing antibodies (Ab1) of defined specificity were sought. In this regard, LAFON et al. (1983) recently presented an operational map delineating at least three major antigenic sites on the CVS

Table 1. Infectious disease models in which Ab2 modulates specific immunity

Antigen	Ab1	Ab2	Characteristic of Ab2 and/or result of Ab2 immunization	References
Bacterial Levan	mAb to $\beta 2 \rightarrow 6$ fructosan	Monoclonal (paratope specific)	Administration at birth activates Id^+ B cells	RUBENSTEIN et al. 1983
			Maturation to Id^+, Ag^+ cells requires antigen challenge or exposure to monoclonal Ab2	BONA et al. 1984
E. coli K13	mAb to P_s capsule	Monoclonal	Primes neonates to homologous bacterial challenge	STEIN and SÖDERSTRÖM 1984
Hepatitis B	Human polyclonal anti-HBsAG	Rabbit polyclonal anti-CRI (paratope specific)	Induction of Id^+, Ag^+, Ab3	KENNEDY and DREESMAN 1984
			Booster response of Id^+, Ag^+, Ab3 after HBsAg exposure	KENNEDY et al. 1983
	mAb to HBsAg	Monoclonal (internal image)	Reactivity of Ab2 with anti-HBsAg from diverse species	THANAVALA et al. 1984
Herpes simplex Type 2	mAb to HSV-2	Rabbit polyclonal (paratope specific)	Decreases survival time of mice challenged with virus	KENNEDY et al. 1985
Micrococcus and tobacco mosaic virus	Rabbit polyclonal	Rabbit polyclonal	Induction of Ag^+ Ab3	WIKLER et al. 1979
			Ab4 recognizes determinants shared between Ab1 and Ab3	URBAIN et al. 1983
Polio virus	mAb to neutralizing epitope	Monoclonal (paratope specific)	Induction of Id^+, Ag^+, Ab3	UYTDE HAAG and OSTERHAUS 1985
Rabies virus	mAb to G protein	Rabbit polyclonal (paratope specific)	Induction of low levels of VN antibodies	
			Booster effect on VN antibodies upon exposure to vaccine	REAGAN et al. 1983
			Dose-dependent modulation of protection	REAGAN et al. 1985
Reovirus	Rabbit polyclonal, or mAb to Sigma 1 protein	Rabbit polyclonal mAb (internal image)	Binding of Ab2 to T cell thymoma and cultured neurons	NEPOM et al. 1982 NOSEWORTHY et al. 1983
			Elicit a T cell-dependent DTH and cytotoxic response	SHARPE et al. 1984
			Induction of reovirus-specific VN antibody	GREENE 1985

Table 1 (continued)

Antigen	Ab1	Ab2	Characteristic of Ab2 and/or result of Ab2 immunization	References
Sendai virus	Murine T helper cell line	Monoclonal (internal image)	Induction of specific cytotoxic T cells	ERTL and FINBERG 1984
			Induction of Id^+, Ag^+, Ab3 and booster response in Ab3 mice upon virus exposure	
			Protection against virus challenge	
Streptococcus pneumoniae	mAb to TEPC 15	Monoclonal (paratope specific)	Induction of Ag^+, Ab3	MCNAMARA et al. 1984
			Increased resistance of Ab3 animals to bacterial challenge	
Trypanosoma rhodesiense	mAb to surface glycoproteins	Allotypic	Induction of Id^+, Ag^-, Ab^3	SACKS et al. 1983
			Protective immunity with Ab3-7H11	SACKS 1983
			Restriction of Ab3 to Igh-C^a gene	SACKS et al. 1983
Venezuelan equine encephalomyelitis virus	mAb to gP56	Rabbit polyclonal (paratope specific)	Induction of Ag^+, Ab3	ROEHRIG et al. 1985

strain rabies virus G protein for type-specific VN antibody (Fig. 1). Five anti-G protein mAb, representing nonoverlapping epitope specificities, were selected from a large panel of hybridomas to produce a polyclonal Ab2 in rabbits. Four of the anti-G protein mAb (509-6, 507-1, 101-1 and 719-3) strongly neutralized virus infectivity and targeted the three known antigenic sites. Each mAb reacted broadly with other laboratory strains of rabies virus as well as with natural isolates, suggesting a recognition of major neutralizing sites on rabies virus. One other anti-G protein mAb, 1104-2, was chosen as a poorly neutralizing antibody which binds to a topographically uncharacterized epitope. Each mAb was purified from ascites fluid using protein A Sepharose chromotography and a hyperimmune serum prepared in rabbits (REAGAN et al. 1983). Individual rabbit Ab2 preparations were extensively absorbed against a purified mAb bearing the identical isotype but unrelated specificity (anti-rabies virus nucleocapsid mAb 515-3 or 389-1, depending on the isotype) in order to eliminate antibodies to constant region determinants. As measured in a radioimmunoassay (RIA), the immunoglobulin fraction which failed to bind to the isotype column con-

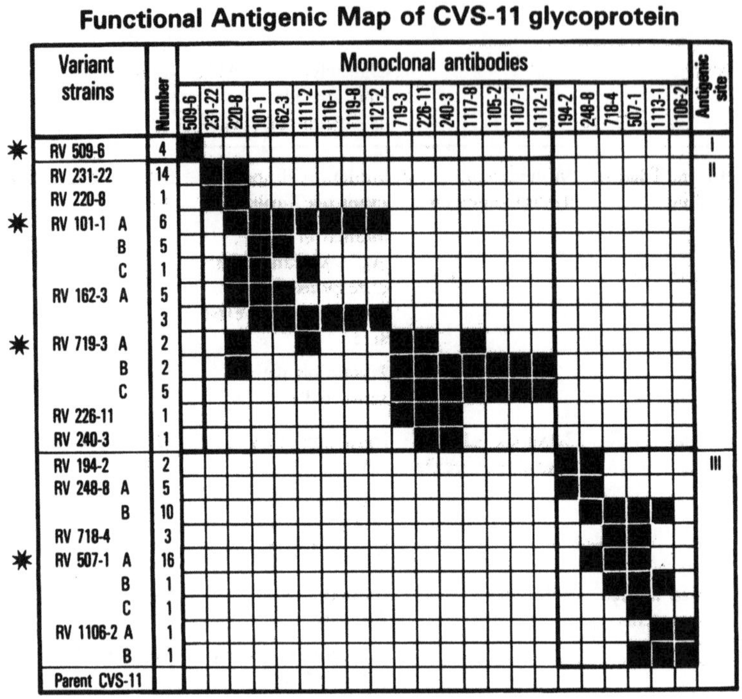

Fig. 1. Functional antigenic map of CVS-11 glycoprotein. Neutralization-resistant variants (RV) were selected from parental virus stock by growth of virus in the presence of individual mAb, then tested for susceptibility (□) or resistance to neutralization by the mAb used for selection of the variant and by other mAb in the panel (■). The notation (*) to the left of the map denotes the mAbs chosen for production of Ab2 (LAFON et al. 1983)

tained the Ab2 specificity. This fraction was subsequently purified using protein A Sepharose chromatography. A cross titration of each Ab2 preparation on homologous and heterologous Ab1 revealed no major cross-reactive Id among the Ab1 chosen in these studies but rather a high specificity for homologous Ab1–Ab2 binding.

To test the relative proportions of each Ab2 response to framework or paratope determinants of its respective Ab1, a competition RIA was devised to test the ability of rabies virus G protein to prevent the interaction of Ab1 and Ab2 (Fig. 2). A soluble form of G protein, designated G_s, purified from virion-depleted culture fluid using affinity chromatography (DIETZSCHOLD et al. 1983), was used as the competing antigen. The binding of Ab2-509-6, Ab2-507-1 and Ab2-1104-2 to their corresponding Ab1 could be partially inhibited by G_s. The maximum inhibition varied from 20% to 50% with as much as 6 μg G_s per ml, but at least a 15% reduction was observed, with as little as 0.75 μg G_s per ml. Because each Ab2 preparation was polyclonal, the inability to obtain total Id inhibition was expected and clearly demonstrated that both framework and paratope specificities were present in three of the Ab2 sera. No inhibition

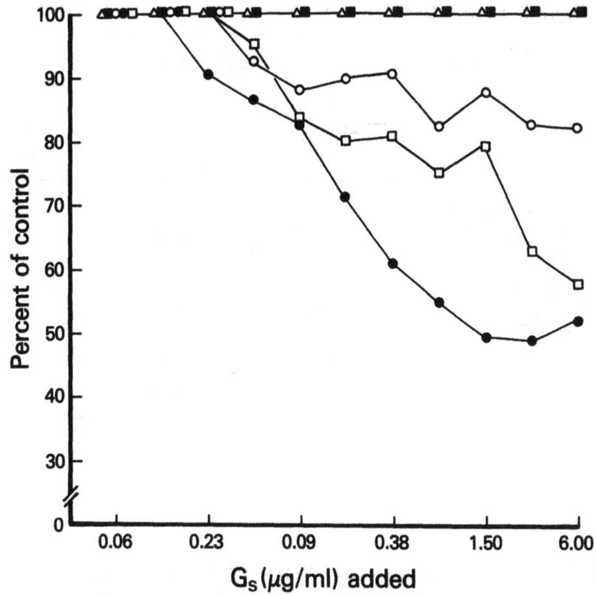

Fig. 2. Competition by G_s for interaction of Ab2 with homologous Ab1. Pretitrated levels of mAb were incubated with serial dilutions of G_s for 1 h before addition of a standardized Ab2 dilution. The amount of Ab2 bound was determined using a ^{125}I-labeled goat antirabbit antibody probe. △, Ab2-719-3; ■, Ab2-101-1; ●, Ab2-509-6; ○, Ab2-507-1; □, Ab2-1104-2 (Reagan et al. 1983)

by G_s was noted for Ab2-101-1 or Ab2-719-3, suggesting the absence or only a minor population of paratope-specific antibodies in these preparations.

3.2 Induction of Virus Neutralizing Antibody by Ab2

To test whether a functional Ab3 could be induced, mice were immunized with individual protein A Sepharose-purified Ab2 preparations (40 μg per mouse). Since Id manipulation will have to be applicable to outbred populations in order to be of use as a vaccine, ICR randomly bred mice were used. The derived sera were reacted with native rabies virus G protein in a neutralization assay (Table 2). Low but specific VN antibody titers were present against ERA strain rabies virus with Ab3-509-6 and Ab3-1104-2. The three other Ab3 preparations as well as mouse serum against normal rabbit IgG failed to neutralize virus. It should be noted that the Ab2 preparations themselves were devoid of detectable VN activity. The specificity of the VN response was evident by the restriction of neutralization to the 509-6 epitope. Only viruses which possessed a 509-6 epitope (ERA parent, CVS parent, and ERA variant virus RV 194-2) were neutralized by Ab3-509-6. Another ERA variant virus, RV 509-6, which is identical to the ERA strain except for the absence of the 509-6 epitope, was not neutralized. Thus, Ab2-509-6 induced a response in outbred mice with

Table 2. Neutralization of rabies virus strains by Ab3 (modified from REAGAN et al. 1983)

Mouse antiserum to	Virus neutralizing titer[a]			
	Strain ERA			Strain CVS
	Parent	RV509-6	RV194-2	Parent
Normal rabbit IgG	<4	<4	<4	<4
Ab2 509-6	32	<4	64	64
Ab2 101-1	<4	ND	ND	ND
Ab2 719-3	<4	ND	ND	ND
Ab2 507-1	<4	ND	ND	ND
Ab2 1104-2	128	<4	128	64

[a] The reciprocal of the highest dilution of serum capable of reducing the number of infected cells by 50% in a modified rapid fluorescence focus inhibition test was taken as the neutralization titer (REAGAN et al. 1983)
ND, not done

the exact specificity as Ab1-509-6. This was not the case with Ab3-1104-2. Ab1-1104-2 showed neither Id cross-reactivity nor an epitope specificity with Ab1-509-6. Yet Ab3-1104-2 showed the same neutralization pattern as Ab3-509-6, which suggests that Ab2-1104-2 induced multispecific antibodies, also observed in other systems (METZGER et al. 1981; KRIEGER et al. 1983). However, mice immunized in this way did not survive a lethal intracerebral challenge with rabies strain CVS. An important point to stress is that the epitope recognized by each Id mAb used in this study was nonlinear (i.e., any denaturation of the G protein destroyed reactivity of the Ab1 in an RIA). Therefore, the ability of two separate Ab2 to induce Ab3 populations with an affinity for such conformationally dependent epitopes, emphasizes a potential for anti-Id vaccination so far lacking in our efforts with synthetic peptides or G protein cleavage fragments (KOPROWSKI et al. 1985; DIETZSCHOLD et al. 1982; WUNNER et al. 1983).

3.3 Booster Effect in Animals Pretreated with Ab2

Ab2 activation of B cells has been manifested as either an elaboration of Id-positive antibody prior to antigen exposure or an enhancement of Id-bearing antibody with postexposure to antigen (SACKS et al. 1983). In order to determine the significance of the low VN antibodies induced by Ab2 against rabies virus as described above, mice were preimmunized with a mixture of the three paratope-related Ab2 (Ab2-509-6, Ab2-507-1, and Ab2-1104-2) 3 weeks prior to the injection of a pretitrated amount of rabies vaccine (KOPROWSKI et al. 1985). Sera were tested for VN antibody immediately prior to and 7 days after vaccine administration. As shown in Table 3, animals pretreated with PBS alone demonstrate a low VN titer indicative of a primary immune response to rabies vaccine. Mice pretreated with the Ab2 mixture show a booster response to rabies vaccine

Table 3. Effect of preimmunization with Ab2 on virus neutralizing antibodies elicited by rabies virus vaccination (KOPROWSKI et al. 1985)[a]

Preimmunized with	Rabies virus neutralizing titer (range) at *n* days postvaccination	
	0	7
Phosphate-buffered saline (PBS)	<10	20
Rabies virus vaccine	20	320–640
Ab2 (mixture)	<10-10	160–320

[a] C57BL/6 mice (in groups of five) were preimmunized with 3 subcutaneous injections of an Ab2 mixture (Ab2-509-6, Ab2-507-1, and Ab2-1104-2; 50 μg each), rabies virus vaccine diluted to 0.04 IU in PBS. Animals were bled prior to and 7 days after vaccination and tested for virus neutralization activity

with VN antibody titers approaching those of animals preimmunized with vaccine. We have concluded that, as with a number of other systems (CAZENAVE 1977; URBAIN et al. 1977; BONA et al. 1981; KENNEDY et al. 1983), Ab2 pretreatment is most effective in sensitizing the immune system to subsequent antigen exposure.

3.4 Ab2 Modulation of Survival from Rabies Virus Challenge

In order to analyze the effect of this regimen on survival of mice against an intracerebral challenge of rabies virus, ICR mice were preimmunized with either PBS, rabies virus vaccine, or Ab2 (individually and in a mixture) 14 days prior to an injection of rabies vaccine diluted to 0.04 IU. All injections were administered i.p. as a soluble preparation in PBS; 7 days later, all animals received an intracerebral challenge with 10 LD_{50} of rabies virus strain CVS and survival up to 3 weeks postinfection was monitored (Table 4). From these data it is apparent that the intracerebral challenge was sufficient to kill all nonimmune animals, while a single dose of vaccine provided 1 week prior to challenge was sufficient to protect only 2/10 animals. In comparison with the animals receiving two doses of vaccine (100% protection), the individual Ab2 pretreatment (50 μg per mouse) did not provide any striking evidence for protection, although the mixture of Ab2 did increase survival. However, in a concurrent experiment, dilutions of Ab2 509-6 ranging from 50 μg to 0.005 μg were provided to animals in the same immunization schedule as above, and an optimal effect on survival was noted at 5 μg (Fig. 3). This protective effect could be reduced with increasing dilution to a point equivalent to that of vaccine alone. Thus, as has been noted by others (EICHMANN 1974; KENNEDY and DREESMAN 1984), Ab2 manipulation of specific immunity is dose dependent.

Theories to explain the induction of specific, functional Ab3 have focused largely on two mechanisms. As in standard antigen processing (Fig. 4, left), B cells committed to recognition of defined epitopes on the rabies G protein

Table 4. Effect of preimmunization with soluble Ab2 plus vaccine on survival against intracerebral challenge with rabies virus (REAGAN et al. 1985)[a]

Preimmunization	Amount	Vaccine	Survival (%)
PBS	–	–	0
PBS	–	+	20
Vaccine	0.04 IU	+	100
Ab2 509-6	50 μg	+	40
Ab2 507-1	50 μg	+	30
Ab2 1104-2	50 μg	+	50
Ab2 (mixture)	50 μg each	+	78 ($n=9$)

[a] ICR mice (in groups of 10) were preimmunized by one i.p. injection (0.5 ml) with Ab2, or rabies virus vaccine (0.04 IU) in PBS, 2 weeks prior to the i.p. injection of vaccine (0.04 IU) in PBS. One week later, animals received 10 LD_{50} of rabies virus intracerebrally

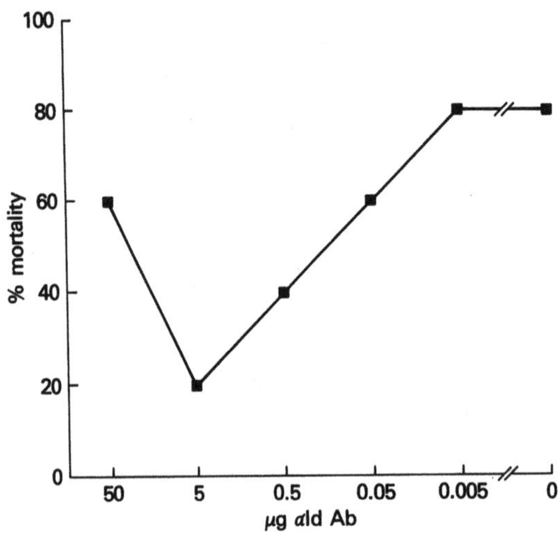

Fig. 3. Effect of dosage of Ab2-509-6 upon survival of ICR mice to lethal intracerebral challenge with rabies virus. Administration of Ab2 and rabies vaccine are described in Table 4

are stimulated after exposure to the relevant conformation. The antibodies produced are defined by their paratopic specificity as well as Id markers (both public and private). When Ab2 which possess "internal images" are generated (JERNE 1974), they are expected to represent conformational homologues of the epitopes recognized by their respective Ab1 (Fig. 4, center). Thus, upon injection of these Ab2 into animals, a stimulation of the same elements of the immune system (both cellular and humoral) as in standard antigen processing would be noted. The induced populations of Ab3 would be expected to possess a restricted paratopic specificity, but be idiotopically diverse (NISONOFF and LAMOYI 1981). According to our data, it might seem that Ab2-509-6 acts as an internal image because it induced Ab3 with an identical specificity as Ab1-509-6. Yet, the B-cell population that expresses Ab3 in response to Ab2-1104-2 also has the specificity of Ab1-509-6, thus indicating that the Ab3 was not equivalent to Ab1. This allows consideration of a second mechanism suggest-

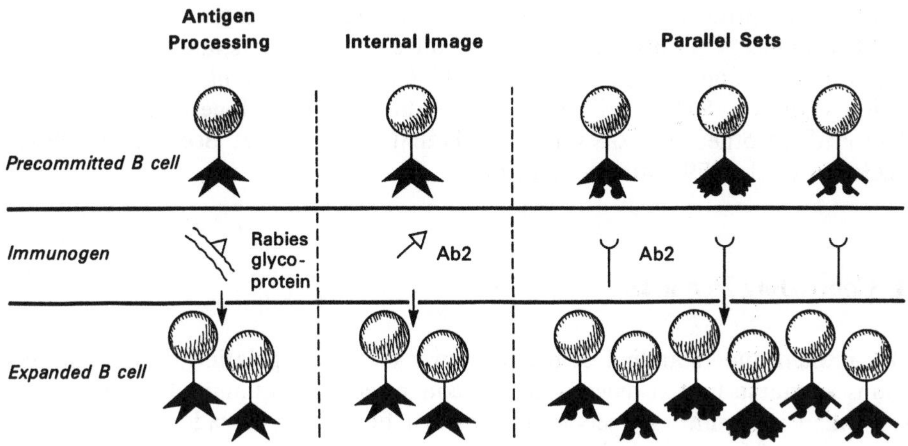

Fig. 4. Models for the stimulation of idiotype-bearing, antigen-binding antibodies by Ab2. Antigen processing, internal image, and parallel sets are described the text (REAGAN et al. 1985)

ing that Ab2 may be stimulating a variety of B cells in nonspecific or even specific parallel sets (Fig. 4, right), all of which share an Id marker yet represent divergent antigen specificity. Immunization with such Ab2 has stimulated largely Id-positive, nonantigen binding molecules (see review, SACKS et al. 1983) with the population of Id-positive, antigen-binding antibodies generally being minor or undetectable and only revealed upon subsequent antigen exposure. Although the induction of cross-reactive Id has been successful in modulating immunity in a number of systems, it is generally felt that such a mechanism places restraints upon the usefulness of Ab2 vaccines since one desires an antibody response of known paratopic specificity. Parallel sets, however, (Fig. 4, right) would result in an Id restriction but with paratopic diversity. In addition, cross-reactive Id are frequently strain-specific, linked to particular heavy-chain allotype genes (PRIMI et al. 1981; SACKS and SHER 1983; BLUESTONE et al. 1981; EPSTEIN et al. 1982; TAKEMORI et al. 1982).

In order to determine which mechamism was operational in the rabies virus sysem, a number of experiments were performed focusing on Ab2-509-6. These may be summarized as follows:

1. The Id marker expressed on Ab1-509-6 is not expressed in human sera from individuals immunized with rabies vaccine nor in rabbits hyperimmunized with native rabies virus G protein, but it is present in the serum of BALB/c mice immunized with vaccine. This suggests that the 509-6 Id marker is not interspecies cross-reactive, but rather indigenous to BALB/mice – a fact which makes it an unlikely candidate for an internal image (NISONOFF and LAMOYI 1981).

2. Immunization of BALB/c mice with individual paratope-specific Ab2 using the regimen previously described (Table 4) led to significant protection from intracerebral challenge in each case. In particular, Ab2-507-1 (in conjunction

with vaccine) protected 100% of animals. This suggests that the failure of Ab2-507-1 to induce VN antibody in outbred mice (Table 2) was due to an absence of relevant V sequences in the genome of ICR mice. It is also likely that immunization with Ab2-509-6 and Ab2-1104-2 induces "silent clones," as has been described by others (CAZENAVE 1977; URBAIN et al. 1977; BONA et al. 1981; LEGUERN et al. 1979; MILLER et al. 1982).

4 Concluding Remarks

Our present understanding of the Id network in the regulation of immune responses suggests the existence of an interconnected series of mutually suppressive (or enhancing) antibody sets which form a functional cascade (JERNE 1976). A number of hapten and infectious disease model systems (see Table 1) have clearly demonstrated that Ab2 are potent modulators of immune responses and that it is attractive to demonstrate a role for such reagents in antigen-independent immunization. The evidence described herein suggests an ability to immunize mice against rabies virus using a polyclonal Ab2, resulting in the appearance of VN antibody to a defined, nonlinear epitope. When administered in conjunction with a minute amount of standard rabies vaccine, increased resistance to a lethal intracerebral challenge of rabies virus is observed. However, the fact that only low VN antibody is induced by Ab2 alone suggests that Ab2 would be more appropriate as a vaccine extender. Although promising, the protection afforded by this regimen is significantly less than that seen with G protein expressed by recombinant technology (WIKTOR et al. 1984; KOPROWSKI et al. 1985). This points out a number of issues to be addressed before Ab2 can be considered as a realistic alternative to conventional vaccination. Certainly one concern lies in the variety of effects one sees with Ab2 treatment with both enhancement and suppression of immune responses noted. This may be largely due to the conditions of Ab2 therapy employed in which the effects of dosage (EICHMANN 1974; KENNEDY and DREESMAN 1984), aggregation of Ab2 (KENNEDY and DREESMAN 1984) and perhaps Ab2 isotype (EICHMANN 1974) are important factors for consideration. The nature of the Id marker to be targeted is also critical. For instance, our data suggests that the Ab2-509-6, which defines antibodies with site I specificity on the rabies G protein, is not interspecies cross-reactive. Our challenge will remain to identify another Id marker which will function either in a species cross-reactive fashion or, preferably, as an internal image of the G protein epitope in order to provide the broadest possible utility. When provided as a monoclonal Ab2 reagent, a homogeneous response may be elicited which allows adequate immunization without subsequent antigen boost.

Until obvious problems (i.e., rapid clearance and immune complex formation) associated with the introduction of a foreign immunoglobulin are resolved, we might explore other avenues for Ab2 use. For instance, is there a possibility for "internal image" Ab2 to be used as an immunodiagnostic for infectious agents which are difficult to propagate? Is there a possibility of bypassing the problems of injecting whole immunoglobulin by determining the sequence of critical V determinants with regulatory potential on monoclonal Ab2 and preparing synthetic peptides for injection? Such an approach has been successful in the α(1–3) epitope of dextran B1355 in which monoclonal Ab1 were sequenced and heavy-chain V identified which possess cross-reactive Id (Clevinger et al. 1980). McMillan et al. (1983) then produced synthetic oligopeptides directed against such Id, which were effective in eliciting a specific Ab2. Such studies, combining the technologies of hybridoma formation, amino acid sequencing and oligopeptide synthesis, have provided insight into the molecular biology of both the immunoglobulin molecule and idiotypy and suggest a more refined methodology for modulating immune responses.

References

Anilionis A, Wunner WH, Curtis PJ (1981) Structure of the glycoprotein gene in rabies virus. Nature 284:275–278

Bluestone JA, Epstein SL, Ozato K, Sharrow SO, Sachs DH (1981) Anti-idiotypes to monoclonal anti-H-2 antibodies. II. Expression of anti-H-2K^K idiotypes on antibodies induced by anti-idiotype on H-2K^K antigen. J Exp Med 154:1305–1318

Bona C, Lieberman R, Chien CC, Mond J, House S, Green I, Paul WE (1978) Immune response to levan. I. Kinetics and ontogeny of anti-levan and anti-inulin antibody response and of expression of cross-reactive idiotype. J Immunol 120:1436–1442

Bona C, Heber-Katz E, Paul WE (1981) Idiotype-anti-idiotype regulation. I. Immunization with a levan-binding myeloma protein leads to the appearance of auto-anti-antibodies and to the activation of silent clones. J Exp Med 153:951–967

Bona CA, Goldberg B, Rubinstein LJ (1984) Regulatory idiotypes, parallel sets, and internal image of the antigen within the polyfructosan A48 idiotype network. Ann Immunol (Paris) 135C:107–115

Bottomly K, Mathieson BJ, Mesier DE (1978) Anti-idiotype induced regulation of helper cell function for the response to phosphorylcholine in adult BALB/c mice. J Exp Med 148:1216–1227

Cazenave PA (1977) Idiotypic-anti-idiotypic regulation of antibody synthesis in rabbits. Proc Natl Acad Sci USA 74:5122–5125

Clevinger B, Schilling J, Hood L, Davie JM (1980) Structural correlates of cross-reactive and individual idiotypic determinants on murine antibodies to alpha (1→3) dextran. J Exp Med 151:1059–1070

Cox JH, Dietzschold B, Schneider LG (1977) Rabies virus glycoprotein. II. Biological and serological characterization. Infect Immun 16:754–759

Cozenza H, Kohler H (1972) Specific suppression of the antibody response by antibodies to receptors. Proc Natl Acad Sci USA 69:2701–2705

Dietzschold B, Wiktor TJ, Macfarlan R, Varrichio A (1982) Antigenic structure of rabies virus glycoprotein: ordering and immunological characterization of the large CNBr cleavage fragments. J Virol 44:595–602

Dietzschold B, Wiktor TJ, Wunner WH, Varrichio A (1983a) Chemical and immunological analysis of the rabies soluble glycoprotein. Virology 124:330–337

Dietzschold B, Wunner WH, Wiktor TJ, Lopes AD, Lafon M, Smith CL, Koprowski H (1983b) Characterization of an antigenic determinant of the glycoprotein that correlates with pathogenicity of rabies virus. Proc Natl Acad Sci USA 80:70–74

Eichmann K (1974) Idiotype suppression. I. Influence of the dose and of the effector functions of anti-idiotypic antibody on the production of an idiotype. Eur J Immunol 4:296–302

Epstein SL, Masakowski VR, Sharrow SO, Bluestone JA, Ozato K, Sachs DH (1982) Idiotypes of anti-IA antibodies. II. Effects of in vivo treatment with xenogeneic anti-idiotype. J Immunol 129:1545–1552

Ertl HC, Finberg RW (1984) Sendai virus specific T cell clones: induction of cytolytic T cells by an anti-idiotypic antibody directed against a helper T cell clone. Proc Natl Acad Sci USA 81:2850–2854

Frischnecht H, Binz H, Wigzell H (1978) Induction of specific transplantation immune reactions using anti-idiotypic antibodies. J Exp Med 147:500–514

Greene MI (1985) Anti-idiotypic immunoglobulin as a vaccine for viral immunity. In: Dreesman G, Kennedy RC (eds) High technology route to viral vaccines. Stem Line, San Antonio (to be published)

Hart DA, Wang AL, Pawlak LL, Nisonoff A (1972) Suppression of idiotypic specificities in adult mice by administration of anti-idiotypic antibody. J Exp Med 135:1293–1300

Jerne NK (1974) Towards a network theory of the immune system. Ann Immunol (Paris) 125C:373–389

Jerne NK (1974–1975) The immune system: a web of V domains. Harvey Lect 70:93–110

Kauffman RS, Noseworthy JH, Nepom JJ, Finberg R, Fields BN, Greene MI (1983) Cell receptors for the mammalian reovirus. Monoclonal anti-idiotypic antibody blocks viral binding to cells. J Immunol 131:2539–2541

Kennedy RC, Dreesman GR (1984) Enhancement of the immune response to hepatitis B surface: in vivo administration of anti-idiotype induces anti-HBs that express a similar idiotype. J Exp Med 159:655–665

Kennedy RC, Adler-Storthz K, Henkel RD, Sanchez Y, Melnick JL, Dreesman GR (1983) Immune response to hepatitis B surface antigen: enhancement by prior injection of antibodies to the idiotype. Science 221:853–855

Kennedy RC, Adler-Storthz K, Burns JW, Henkel RD, Dreesman GR (1984) Antiidiotype modulation of herpes simplex virus infection leading to increase pathogenicity. J Virol 50:951–953

Kieny MP, Lathe R, Drillien R, Spehner D, Skory S, Schmitt D, Wiktor T, Koprowski H, Lecocq JP (1984) Expression of rabies virus glycoprotein from a recombinant vaccinia virus. Nature 312:163–166

Koprowski H, Reagan KJ, Macfarlan RI, Dietzschold B, Wiktor TJ (1985) New generation of rabies vaccines: rabies glycoprotein gene recombinants, anti-idiotypic antibodies, and synthetic peptides. In: Modern approaches to vaccines. Cold Spring Harbor Laboratory, New York p 151–156

Krieger NJ, Pesce AJ, Michael JG (1983) Induction of multi-specific antibodies to bovine serum albumin after production of anti-idiotypic antibodies to an albumin-specific monoclonal antibody. Ann NY Acad Sci 418:305–312

Kunkel HG, Mannik M, Williams RC (1963) Individual antigenic specificity of isolated antibodies. Science 140:1218–1219

Lafon M, Wiktor TJ, Macfarlan RI (1983) Antigenic sites on the CVS rabies virus glycoprotein: analysis with monoclonal antibodies. J Gen Virol 64:843–851

LeGuern C, BenAissa F, Juy O, Mariame B, Buttin G, Cazenave PA (1979) Expression and induction of MOPC-460 idiotypes in different strains of mice. Ann Immunol (Paris) 130C:293–302

Macfarlan RI, Dietzschold B, Wiktor TJ, Kiel M, Houghten R, Lerner RA, Sutcliffe JG, Koprowski H (1984) T cell responses to cleaved rabies virus glycoprotein and to synthetic peptides. J Immunol 133:2748–2752

McMillan S, Seiden MV, Houghten EA, Clevinger B, Davie JM, Lerner RA (1983) Synthetic idiotopes: the third hypervariable region of murine anti-dextran antibodies. Cell 35:859–863

McNamara MK, Ward RE, Kohler H (1984) Monoclonal idiotope vaccine against Streptococcus pneumoniae infection. Science 226:1325–1326

Metzger DW, Furman A, Miller A, Sercarz EE (1981) Idiotypic repertoire of anti-HEL antibodies probed with hybridoma antibodies. Selection after immunisation of an IdX marker common to antibodies of distinct epitope specificity. J Exp Med 154:701–712

Miller GG, Nadler PI, Hodes RJ, Sachs DH (1982) Modification of T cell antinuclease idiotype expression by in vivo administration of antiidiotype. J Exp Med 155:190–200

Nepom JT, Weiner HL, Dichter MA, Tardieu M, Spriggs DR, Gramm CF, Powers ML, Fields BN, Greene MI (1982) Identification of a hemagglutinin-specific idiotype associated with reovirus recognition shared by lymphoid and neural cells. J Exp Med 155:155–167

Nisonoff A, Lamoyi E (1981) Hypothesis. Implications of the Presence of an internal image of the antigen in anti-idiotypic antibodies: possible application to vaccine production. Clin Immunol Immunopathol 21:397–406

Noseworthy JH, Fields BN, Dichter MS, Sobotka C, Pizer E, Perry LL, Nepom JT, Green MI (1983) Cell receptors for the mammalian reovirus. I. Syngeneic monoclonal anti-idiotypic antibody identifies the cell surface receptor for reoviruses. J Immunol 131:2533–2538

Oudin J, Michel M (1963) Une nouvelle forme d'allotypie des globulines γ du sérum de lapin aparement liée à la jonction et à la spécificité anticoyss. C R Soc Biol 257:805–808

Primi D, Juy D, Cazenave PA (1981) Induction and regulation of silent iodiotype clones. Eur J Immunol 11:393–398

Reagan KJ, Wunner WH (1984) Early interactions of rabies virus with cell surface receptors. In: Bishop DHL, Compans RW (eds) Nonsegmented negative strand viruses. Academic, New York, pp 387–392

Reagan KJ, Wunner WH, Wiktor TJ, Koprowski H (1983) Anti-idiotypic antibodies induce neutralizing antibodies to rabies virus glycoprotein. J Virol 48:660–666

Reagan KJ, Wunner WH, Koprowski H (1985) Viral antigen-independent immunization: induction by anti-idiotypoic antibodies of neutralizing antibodies to defined epitopes on rabies virus glycoprotein. In: Dreesman G, Kennedy RC (eds) High technology route to viral vaccines. Stem Line, San Antonio (to be published)

Rodkey LS (1980) Autoregulation of immune responses via idiotypic network interactions. Microbiol Rev 44:631–659

Roehrig JT, Hunt AR, Mathews JH (1985) Identification of anti-idiotype antibodies that mimic the neutralization site of Venezualan equine encephalomyelitis virus. In: Dreesman G, Kennedy RC (eds) High technology route to viral vaccines. (to be published) Stem Line, San Antonio

Rubenstein LJ, Goldberg B, Hiernaux J, Stein KE, Bona CA (1983) Idiotype-antiidiotype regulation. V. The requirement for immunization with antigen or monoclonal antiidiotype for the activation of $\beta 2 \rightarrow 6$ and $\beta 2 \rightarrow 1$ polyfructosan-reactive clones in BALB/c mice treated at birth with minute amounts of anti-A48 idiotype antibodies. J Exp Med 158:1129–1144

Sachs DH (1981) Genetic control of idiotype expression. In: Fougereau M, Dausset J (eds) Immunology 80: Progress in immunology IV. Academic, London, p 478–495

Sacks DL, Sher A (1983) Evidence that anti-idiotype induced immunity to experimental African trypanosomiasis is genetically restricted and requires recognition of combining site-related idiotopes. J Immunol 131:1511–1515

Sacks DL, Esser KM, Sher A (1982) Immunization of mice against African trypanosomiasis using anti-idiotypic antibodies. J Exp Med 155:1108–1119

Sacks DL, Kelsoe GH, Sachs DH (1983) Induction of immune responses with anti-idiotypic antibodies: implications for the induction of protective immunity. Springer Semin Immunopathol 6:79–97

Sharpe RH, Gaulton GN, McDade KK, Fields BN, Green MI (1984) Syngeneic monoclonal antiidiotype can induce cellular immunity to reovirus. J Exp Med 160:1195–1205

Stein KE, Söderström T (1984) Neonatal administration of idiotype or antiidiotype primes for protection against *E. coli* K13 infection in mice. J Exp Med 160:1001–1011

Takemori T, Tesch H, Reth M, Rajewsky K (1982) The immune response against anti-idiotope antibodies. I. Induction of idiotope-bearing antibodies and analysis of the idiotope repertoire. Eur J Immunol 12:1040–1046

Thanvala YM, Bond A, Tedder R, Hay FC, Roitt IM (1985) Monoclonal internal image anti-idiotype antibodies of hepatitis B surface antigen. In: Dreesman G, Kennedy RC (eds) High technology route to virus vaccines. Stem Line, San Antonio (to be published)

Trenkner E, Riblet R (1975) Induction of anti-phosphorylcholine antibody formation by anti-idiotypic antibodies. J Exp Med 142:1121–1132

Urbain J, Wikler M, Franssen JD, Collignon C (1977) Idiotypic regulation of the immune system by the induction of antibodies against anti-idiotypic antibodies. Proc Natl Acad Sci USA 74:5126–5130

Urbain J, Francotte M, Franssen JD, Heirnaux J, Leo O, Moser M, Slaoui M, Urbain-Vansanten G, Van Acker A, Wikler M (1983) From clonal selection to immune networks: induction of silent idiotypes. Ann NY Acad Sci 418:1–8

UytdeHaag FG, Osterhaus AD (1985) Induction of neutralizing antibody in mice against poliovirus type II with monoclonal anti-idiotypic antibody. J Immunol 134:1225–1229

Wikler M, Franssen JD, Collignon C, Leo O, Mariame B, Van De Walle P, DeGroote D, Urbain J (1979) Idiotypic regulation of the immune system. Common idiotypic specificities between idiotypes and antibodies raised against anti-idiotype antibodies in rabbits. J Exp Med 150:184–195

Wiktor TJ, Gyorgy E, Schlumberger MD, Sokol F, Koprowski H (1973) Antigenic properties of rabies virus components. J Immunol 110:269–276

Wiktor TJ, Kamo I, Koprowski H (1974) In vitro stimulation of rabbit lymphocytes after immunization with live and inactivated rabies vaccines. J Immunol 112:2013–2019

Wiktor TJ, Macfarlan RI, Reagan KJ, Dietzschold B, Curtis PJ, Wunner WH, Kieny MP, Lathe R, Lecocq JP, Mackett M, Moss B, Koprowski H (1984) Protection from rabies by a vaccinia virus recombinant containing the rabies virus glycoprotein gene. Proc Natl Acad Sci USA 81:7194–7198

Wunner WH, Dietzschold B, Curtis PJ, Wiktor TJ (1983) Review article: rabies subunit vaccines. J Gen Virol 64:1649–1656

Wunner WH, Reagan KJ, Koprowski H (1984) Characterization of saturable binding sites for rabies virus. J Virol 50:691–697

Vaccines from Monoclonal Anti-Idiotypic Antibody: Poliovirus Infection as a Model

F.G.C.M. UYTDEHAAG and A.D.M.E. OSTERHAUS

1 Introduction

Manipulation of the immune system by "foreign" substances, i.e., the induction of protective immune response against certain pathogens by vaccination has proven to be of great practical value in contributing to the control of infectious diseases in man and animals. In the case of smallpox, it has even led to the eradication of the disease. However, for a number of infectious diseases there are still major problems in the production of effective vaccines. This is partly due to the fact that antigens cannot be generated in sufficient amounts or in the proper immunogenic form. Among the different approaches to overcoming these problems (reviewed by ARNON et al. 1983 and NORRBY 1983) the use of the elements of the immune system itself, i.e., anti-idiotypic antibodies has recently attracted great attention (see e.g. NISONOFF and LAMOYI 1981 and ZOLER 1984). In light of present idiotype (Id) research, it must be considered possible to administer anti-Id antibody (Ab2) exogenously to replace antigen for the induction of specific immune response (see for review, RAJEWSKI and TAKEMORI 1983 and SACKS and SHER 1983).

Several groups, using biologically important models of microbial infection, have now shown that the administration of polyclonal xenogeneic or allogeneic Ab2 to animals can either prime for a protective antibody response upon subsequent antigen exposure, or can induce neutralizing antibodies in animals in absence of antigen (SACKS et al. 1981; 1983; KENNEDY et al. 1982, 1983a, b; KENNEDY and DREESMAN 1983, 1984; REAGAN et al. 1983). In principle, apart from obvious theoretical implications, polyclonal xenogeneic or allogeneic Ab2 preparations for the development of a vaccine have several major practical disadvantages. (a) It would be very difficult to establish production of polyclonal

The National Institute of Public Health and Environmental Hygiene, P.O. Box 1, NL-3720 BA Bilthoven

Current Topics in Microbiology and Immunology, Vol. 119

Ab2 preparations of high consistency and identity. (b) A polyclonal Ab2 preparation would contain a heterogeneous population of anti-Id specificities where only a small fraction induces Id-bearing molecules which will also bind antigen. The rapidly progressing developments in hybridoma technology (for review, see e.g., OSTERHAUS and UYTDEHAAG 1985 have raised the possibility for large-scale production of homogeneous Id-bearing and anti-Id antibody preparations of high consistency and purity. A monoclonal Ab2 detecting an interspecies crossreactive idiotope (CRI) associated with the immune response against, for example, a neutralization domain of a viral epitope or possessing an "internal image" of such an epitope would represent the ideal anti-Id vaccine. The induction of antiviral T-cell mediated immunity by monoclonal Ab2 and monoclonal anti-T cell idiotope antibodies has also been reported (NOSEWORTHY et al. 1983; KAUFFMAN et al. 1983; SHARPE et al. 1984; ERTL and FINBERG 1984).

In the present paper, we will consider the potential of monoclonal Ab2 for the induction of protective immune response against poliovirus infection. Endemic poliomyelitis has been exterminated from several industrialized countries with either oral attenuated poliovirus or inactivated poliovirus vaccines. However, also in these countries poliomyelitis remains a permanent threat as was recently shown by an outbreak of the disease caused by poliovirus type III in Finland (ANONYMOUS (1985). In the developing world, this disease is still a major problem, and there is urgent need for an effective vaccine that can be produced at low cost for worldwide distribution, perhaps even with the final goal of complete eradication of the disease. For this purpose, a vaccine not containing live virus would be the most suitable candidate. The presently available inactivated polio vaccines will not fill this gap since they are produced in primary or subcultivated monkey-kidney cells, which cannot be produced in sufficient amounts at sufficiently low cost. Among the recent developments in this field is the evaluation of similar vaccines produced in continuous monkey-kidney cell lines (BEALE 1980; VON SEEFRIED and CHUN 1980; HORODNICEANU et al. 1980). Although this approach will certainly expand the potential of inactivated polio vaccines on a quantitative basis, the discussion about the safety of these vaccines with regard to the tumorogenous potential of the cells used for production has not been fully evaluated (PROCEEDINGS OF THE WORKSHOP ON ABNORMAL CELLS, NEW PRODUCTS AND RISKS 1984). At least, this will lead to the requirement of extra purification steps, to demonstrate the absence of cellular nucleic acids. Developments in the field of the generation of a recombinant DNA or even a synthetic vaccine against poliovirus have been very encouraging, although especially the importance of conformational antigenic determinants may seriously hamper this approach (EMINI et al. 1983; EVANS et al. 1983; MINOR et al. 1983; B. Wieringa, personal communication).

In a first approach to test the feasibility of a monoclonal Ab2 as an antiviral vaccine we choose the poliovirus system mainly because of the assumed limited number of neutralization-inducing epitopes on polioviruses (FERGUSSON et al. 1984; EVANS et al. 1983). This would enable us in principle to use a limited panel of monoclonal Ab2 as a vaccine. Furthermore, humoral immunity is probably the major component of the protective immune response against infection with poliovirus. The third and perhaps most practical reason to choose this model is that a protective immune response in mice may be elicited by one single inoculation of one monoclonal neutralizing antipoliovirus antibody (Osterhaus, personal communication). Finally, a recently developed in vitro system for the induction and measurement of a poliovirus-specific neutralizing antibody response using human peripheral blood lymphocytes (UYTDEHAAG et al. 1985a) will offer the possibility to study at least some aspects of the Ab2-induced humoral immune response in man.

2 Generation of Antipoliovirus Type II Neutralizing Antibody by Monoclonal Ab2

Three monoclonal Ab1 were selected from a panel of monoclonal antibodies with specificity for neutralization-inducing epitopes of the MEF_1 strain of poliovirus type II (OSTERHAUS et al. 1981a, b, 1983). Selection was based on the broad reactivities of the monoclonal Ab1 against various poliovirus type II strains, as tested in a microvirus neutralization test (Table 1). Considering these observations, as well as the facts that the three monoclonal Ab1 derived from two different fusions and the binding of one monoclonal Ab1, 1-10C9E8, to the virus could be inhibited by the other two (Fig. 1), it was deduced that these three Ab1 define a major (and probably the same) neutralization-inducing epitope on poliovirus type II strain MEF_1.

Ab1 1-10C9E8 (IgG2a κ) purified from ascitic fluid using protein A-sepharose affinity chromatography, was subsequently used to raise syngeneic mono-

Table 1. Inter- and intratypic micro-VN titers of antipoliovirus monoclonl antibodies (Ab1)

Polioviruses	Monoclonal antibodies		
	81/21 MEF 1-10C9E8	81/21 MEF 4-15D4E8	81/10 MEF 11E7
Type II			
MEF_1	781.250	781.250	781.250
Sabin	156.250	156.250	6.250
SL 188/4/3	3.906.250	3.906.250	156.250
NSL 81-4789	781.250	≥3.906.250	156.250
NSL 77-728	781.250	≥3.906.250	156.250
NSL 2188	781.250	≥3.906.250	156.250
NSL 77-686	3.906.250	≥3.906.250	156.250
NSL 64-80	3.906.250	3.906.250	156.250
Type I			
Mahoney	<10	<10	<10
Sabin	<10	<10	<10
Type III			
Saukett	<10	<10	<10
Sabin	<10	<10	<10

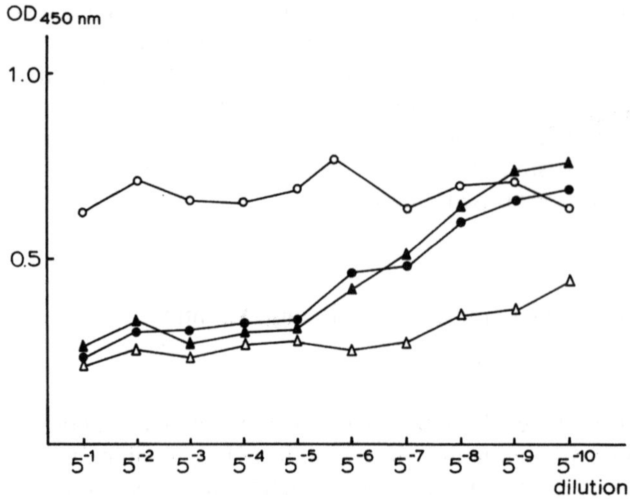

Fig. 1. Inhibition of the binding of peroxidase conjugated Ab1 to 1-10C9E8 to plate-bound poliovirus type II by the homologous and two other heterologous monoclonal antipoliovirus type II antibody preparations. Samples Ab1 1-10C9E8 (▲), Ab1 11E7 (●), Ab1 4-15D4D8 (△), and control BALB/c IgG2a κ (○) were reacted with plate-bound poliovirus, as described in text. Control BALB/c IgG2a κ used as inhibitor resulted in OD 450 nm values comparable with values obtained with PBS-Tween 0.05% as inhibitor

clonal Ab2. One out of five clones obtained in three independent fusions showed strong reactivity for Ab1 1-10C9E8 as tested in an idiotope-crosslinking ELISA (see legend to Fig. 2). This Ab2 clone, 2-17C3scc, was recloned using manual isolation of single cells and monoclonal Ab2 2-17C3scc (IgGl κ) was isolated from the hybridoma supernatant fluid using protein A-sepharose affinity chromatography. Ab2 2-17C3scc showed only binding affinity for Ab1 1-10C9E8 and not for a control BALB/c IgG2a κ monoclonal antibody in an Ab1-crosslinking ELISA. In addition, no binding of a control monoclonal IgG1 κ to Ab1 1-10C9E8 or to the control IgG2a κ was found (Fig. 2A).

The idiotope on Ab1, 1-10C9E8, as defined by Ab2, 2-17C3scc, was found to be closely associated with its paratope. The binding of horseradish peroxidase labeled Ab2 2-17C3scc to Ab1 1-10C9E8 f(ab′)2 could completely be inhibited by poliovirus type II, but not poliovirus type I or type III (Fig. 2B).

A series of competitive inhibition studies were performed to characterize the idiotope defined by Ab2 2-17C3scc. First, either monoclonal neutralizing Ab1 or non neutralizing Ab1 against the different types of poliovirus were tested for inhibition of Ab1 1-10C9E8 idiotope crosslinking by Ab2 2-17C3scc in ELISA. To that end amounts of Ab2 2-17C3scc, at a dilution giving 50% binding in Ab1 1-10C9E8 crosslinking ELISA, were incubated overnight with equal amounts of serial dilutions of the various monoclonal Ab1. Then, the remaining anti-Ab1 1-10C9E8 activity in Ab2 was determined in the Ab1 1-10C9E8 crosslinking ELISA. A dose-related inhibition in Ab1 1-10C9E8 crosslinking was observed by preincubation of Ab2 2-17C3scc with Ab1 1-10C9E8, and with two other monoclonal antipoliovirus type II monoclonal antibodies,

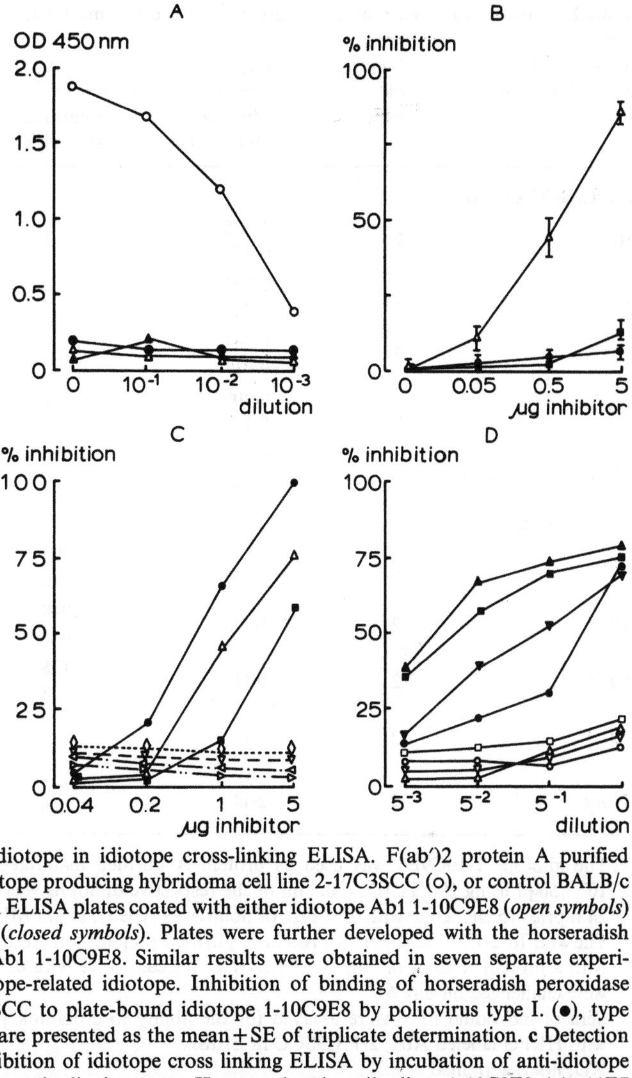

Fig. 2. A Detection of anti-idiotope in idiotope cross-linking ELISA. F(ab′)2 protein A purified supernatant fluids of anti-idiotope producing hybridoma cell line 2-17C3SCC (○), or control BALB/c IgG2a κ (△) were incubated on ELISA plates coated with either idiotope Ab1 1-10C9E8 (*open symbols*) or control BALB/c IgG2a κ (*closed symbols*). Plates were further developed with the horseradish peroxidase labeled idiotope Ab1 1-10C9E8. Similar results were obtained in seven separate experiments. **b** Detection of paratope-related idiotope. Inhibition of binding of horseradish peroxidase labeled anti-idiotope 2-17C3SCC to plate-bound idiotope 1-10C9E8 by poliovirus type I. (●), type II (△), and type III (■). Data are presented as the mean ± SE of triplicate determination. **c** Detection of crossreactive idiotope. Inhibition of idiotope cross linking ELISA by incubation of anti-idiotope 2-17C3SCC with neutralizing antipoliovirus type II monoclonal antibodies; 1-10C9E8 (●), 11E7 (△), 4-15D4D8 (■) or other monoclonal antibodies; nonneutralizing antipoliovirus type II (◇), neutralizing antipoliovirus type I (◁), neutralizing antipoliovirus type III (▽), and control BALB/c IgG2a κ (▷). Similar results were obtained in three separate experiments. **D** Detection of interspecies crossreactive idiotopes. Inhibition of idiotope crosslinking ELISA by incubation of anti-idiotope 2-17C3 with antipoliovirus type II hyperimmune sera (closed symbols) of rats (▲), guinea pigs (■), mice (●), and humans (▼), or their respective preimmune sera (*open symbols*). One pannel of representative data is given from experiments in which sera were tested from 6 rats, 2 guinea pigs, 12 mice, and 12 human individuals

Ab1 4-15D4E8 and Ab1 11E7, but not with non-neutralizing antipoliovirus type II or neutralizing antipoliovirus type I and III monoclonal antibodies (Fig. 2C). Since Ab1 1-10C9E8 and Ab1 4-15D4E8 were obtained from a hybridization other than Ab1 11E7, these results strongly indicate that the

Table 2. Anti-Id 2-17C3SCC induces expression of neutralizing 1-10C9E8 related Id

Mice inoculated with dilution of	Number of mice	Expression: Idiotype[c] OD 450 nm	Expression: Neutralizing antibody	Survival[a] after challenge days
2-17C3SCC anti-Id[b]				
10^{0}	2	1.635 0.477	32	5 10
10^{-1}	2	0.594 0.931	32	10 11
10^{-2}	2	0.341 0.611	<2	9 12
10^{-3}	2	0.209 0.178	<2	8 7
Normal BALB/c Ig	2	0.107 0.119	<2	9 10
1-10C9E8 id[c]				
10^{0}	2	ND[d]	>4096 >4096	>25
10^{-1}	2	ND	2048 2048	>25
10^{-2}	2	ND	128 128	>25
10^{-3}	2	ND	8 2	15 13
None	2	ND	<2 <2	14 9

[a] The idiotope 1-10C9E8 was used as protein A purified material with a neutralization titer of $10^{4.9}$/50 µl using 100 $TCID_{50}$. 0.5 ml of the respective dilutions was inoculated in individual mice, 24 h prior to challenge with 20 LD_{50} of poliovirus type II, strain MEF1

[b] The anti-Id 2-17C3SCC was a protein-A purified preparation from hybridoma supernatant fluid with a protein content of 5 µg/ml. Mice were injected twice with 0,5 ml anti-Id in saline. Animals were challenged on day 6

[c] Sera of individual mice were assayed for the ability to cross-link plate-bound Ab2, 2-17C3SCC with HRPO labeled Ab2, 2-17C3SCC in ELISA

[d] ND, not determined

1-10C9E8 is an intrastrain crossreactive idiotope. This conclusion was confirmed by the observation that sera from BALB/c mice hyperimmunized with poliovirus type II also inhibited Ab1 1-10C9E8 idiotope crosslinking by Ab2 2-17C3scc, while preimmune sera of the animals failed to inhibit this reaction. Moreover, antipoliovirus type II hyperimmune sera of rats, guinea pigs, and humans tested in neutralization assays, but not their preimmune sera, inhibited Ab1 1-10C9E8 idiotope crosslinking by Ab2 2-17C3scc (Fig. 2D).

These results provide evidence for the expression of the 1-10C9E8 idiotope as a major crossreactive idiotope in association with a neutralizing antibody response towards poliovirus type II in different species.

Because Ab2 2-17C3scc defines (at least operationally) an interspecies crossreactive idiotope, experiments were carried out aiming at the induction of protective immune response against poliovirus in mice by immunization with Ab2 2-17C3scc. BALB/c mice were inoculated i.p. with various dilutions in saline of 5 µg/ml purified Ab1 1-10C9E8 (neutralizing titer, $10^{4.9}$/50 µl using 100 $TCID_{50}$), Ab2 2-17C3scc, or normal BALB/c IgG. Animals injected with Ab1 1-10C9E8 were challenged intracerebrally 24 h later using 20 LD_{50} of a mouse brain adapted MEF_1 poliovirus type II strain. Mice, which expressed neutralizing antibody titers ≥ 128, survived virus challenge, whereas the animals expressing <128 neutralizing antibody titer died within 15 days after challenge as did the nonantibody inoculated control mice. Animals injected with Ab2 2-17C3scc developed a dose-dependent Ab3 antibody response exhibiting 1-10C9E8 idiotope specificity (Table 2). Poliovirus neutralizing antibody were only detected in mice injected with 5 µg and 0.5 µg Ab2 (Table 2). Since Ab2 2-17C3 itself contains no poliovirus neutralizing activity and does not appear to act as a polyclonal activator (unpublished observations), and the mice may be assumed never to be exposed to poliovirus antigen, we concluded from this experiment that Ab2 2-17C3scc acts by expanding 1-10C9E8 crossreactive Id-bearing B-cell clones with specificity for a major neutralization-inducing epitope of poliovirus type II.

3 Discussion

We have demonstrated that immunization of mice with a monoclonal syngeneic Ab2, directed against a neutralizing antipoliovirus type II monoclonal Ab1, without antigen, induces an Ab3 response, at least a part of which consists of a population of antibody that shares two properties with Ab1: (a) the binding of poliovirus type II strain MEF_1 and (b) the binding to Ab2.

Since the idea of using elements of the immune system itself, anti-Id antibodies, to induce immune response against infectious agents was formulated by Nisonoff and Lamoyi (NISONOFF and LAMOYI 1981), a few years after the primordial concept of the Id cascade by Cazenave (CAZENAVE 1977) and Urbain (URBAIN et al. 1977), the potential for the practical use of anti-Id antibodies as vaccines has generally been regarded with scepticism. Among the theoretical and practical drawbacks considered are (a) the possible generation of anti-allotype and/or anti-isotype reactivity, (b) the fact that in some systems only a priming effect can be generated, (c) the chance to generate antibodies of unknown epitope specificity (auto-immune disease?), and (d) the genetic restriction of the induction of immunity in some systems. However, although no final evaluation of these problems can as yet be made, to a certain extent they may be irrelevant, if one aims at the generation of Ab2 which functions by expanding sets of idiotope-positive clones, some of which are of predetermined epitope specificity (priming). Subsequent boostering by the actual epitope will expand only the clones with corresponding epitope specificity, leaving the rest of the sets initially induced by Ab2 unaffected.

The general strategy for the preparation of an Ab2 vaccine has so far been the generation of Ab2-defining interspecies CRI, which induce Ab3-sharing antigen-binding capacity with Ab1. In the present studies Ab2 2-17C3scc defines not only a paratope-related idiotope on Ab1 1-10C9E8, but also idiotopes on two other poliovirus type II neutralizing monoclonal antibodies. Since screening of hyperimmune sera from several species, including man, revealed a striking association between the Ab2 2-17C3scc binding idiotope and poliovirus type II specific antibody, these results strongly indicate that Ab2 2-17C3scc defines an interspecies CRI that constitutes a major component of the immune response to poliovirus type II. The phenomenon of interspecies CRI has been interpreted as the conservation of genes in the germline through extended periods of evolution (JU et al. 1978, 1979; THEZE and MOREAU 1978; CLAFLIN and DAVIE 1974; SOMMÉ et al. 1984; KENNEDY et al. 1983 b and KENNEDY and DREESMAN 1984). We also favor this interpretation to explain the expression of the Ab1 1-10C9E8 interspecies CRI observed in the poliovirus system. The recent development in our laboratory of an in vitro system for the induction of a poliovirus type-specific secondary neutralizing antibody response using human peripheral blood lymphocytes, will hopefully allow us to further investigate the regulation of the expression of the 1-10C9E8 idiotope in an antipoliovirus response (UYTDEHAAG et al. 1985a).

An alternative way to explain CRI follows from the consideration that the antibody repertoire of each individual is complete (COUTINHO 1980 and COUTINHO et al. 1983, 1984). This, as stated by JERNE (1984), implies that the immune system not only has the capability to recognize the universe of "foreign" and self-epitopes but is bound, because of the completeness of its repertoire, to recognize its own elements as well. In other words, within one individual idiotopes can be recognized by the combining sites (paratopes) of other molecules and vice versa. Consequently, as has been postulated by (JERNE (1981 and JERNE et al. 1982), anti-idiotopes are capable of mimicking the universe of external and self epitopes, i.e., they may constitute the internal images of epitopes. Internal images of epitopes are most likely to be found in the population of Ab2. They were termed Ab2 by JERNE (1981 and JERNE et al. 1982), and homobodies by LINDEMAN (LINDEMAN 1973). Thus, the reaction of an Ab2 with CRI present in an Ab1 population can be explained on the basis of an internal image on Ab2 carrying an idiotope that resembles the original epitope and consequently reacts with the paratope of Ab1. The concept of internal image is best illustrated by what is well known as a "classical" example of internal image: Ab2 in the rabbit b6 allotype system (reviewed by CAZENAVE and ROLAND 1984). Other examples of Ab2 that were found to behave like the antigen (i.e., a hormone), even in a functional way, have been reported (reviewed by STROSBERG 1983). Notwithstanding the supportive data and the fact that in light of JERNE's network theory (1974, 1976) the existence of internal images is virtualy inevitable, their existence has been questioned (LINDEMANN 1978) and seriously challenged recently by ROUX et al. (1984) and FRANSSEN and URBAIN (1985).

While until now the applicability of Ab2 vaccines seemed restricted to internal image Ab2 and/or anti-CRI Ab2, new possibilities have recently emerged from experiments using the immunization cascade idea.

The primordial idea of Ab2 vaccines is the Id cascade initially demonstrated by CAZENAVE (1977) and URBAIN et al. (1977). They showed that a rabbit immunized with Ab2 in response to a given antigen can learn to make the same private Id used by another rabbit in response to the same antigen (induction of silent idiotopes). FRANCOTTE and URBAIN (1984) demonstrated that it is possible to use an Ab2 not displaying the properties of an internal image to induce expression of a rabbit "private" Id in mice. They induced Ab3 in BALB/c mice by injecting rabbit Ab2. These Ab2 were directed against a private Id of rabbit Ab1 anti-TMV specific antibody. In the induced Ab3, they observed the appearence of mouse anti-TMV antibodies sharing the Id of Ab1. This occurred in spite of the fact that the mice had never been exposed to TMV and do not normally express the private rabbit Id. These results obtained in the TMV system, indicate that in principle any Ab2 against a private Id expressed in a particular individual in response to AgX, can induce this Id in a crossreactive fashion in another individual or species which normally does not express this Id in response to the same AgX. This illustrates that in contrast to what was generally believed initially – the exclusive role of internal image Ab2 and/or Ab2 defining interspecies CRI – also posibilities to expand the available repertoire by activation of silent clones bearing private idiotopes (or intrastrain CRI, see LEO et al. 1984; MEEK et al. 1984; SIGAL 1982; LUCAS and HENRY 1982; MOSER et al. 1983) should be seriously considered. All these approaches, however, have in common that they induce an Ab3 population that binds the original epitope and in most cases share idiotopes with Ab1. There are three possible mechanisms whereby Ab2 induces protective immunity:

1. Ab2 representing internal images of the original epitopes, inducing Ab3 with similar epitope-binding capacity of Ab1, but for the greater part not sharing the CRI of Ab1.
2. Ab2 defines intrastrain, intraspecies, or interspecies CRI, inducing Ab3-sharing CRI and antigen-binding capacity with Ab1.
3. Ab2 defines a private Id, inducing this Id in Ab3 response in a recurrent fashion (CRI) either intrastrain, intraspecies, or interspecies.

For practical reasons it is obvious that future Ab2 vaccines should be composed of monoclonal antibody reagents. The potential of using such monoclonal Ab2 to induce poliovirus neutralizing antibody has been demonstrated in the present paper. This is the first report on the use of monoclonal Ab2 to be used for vaccine purposes. A similar approach has recently been indicated for the reovirus system by SHARPE et al. (1984). One should, however, bear in mind that such vaccines, because of their unique anti-idiotope specificity will induce antibody of unique epitope specificity. This bears the intrinsic danger of in vivo induction of variant viruses resistant to the immune response on the basis of mutant selection. Therefore, panels of monoclonal Ab2, produced against poly- or monoclonal Ab1, defining different epitopes, should be used. Although the routine production of monoclonal antibodies is still restricted to a small number of animal systems (mouse, rat, and man), this is not likely to constitute a major disadvantage in light of the results of Urbain's group (mentioned above). In addition to this approach, also the direct immortalization of Ab2 producing

B-cell clones after in vivo or in vitro immunization with antigen may be considered.

As is also the case for recombinant DNA and synthetic vaccines, we are confronted with the question of how to present the immunogenic moiety to the immune system. At present it is not known how or whether these structures require a specific way of presentation, although it was shown that a multimeric form of presentation in combination with class I and II antigens was more efficient (KAUFMANN et al. 1983).

The fact that the mice exhibiting virus-neutralizing (VN) antibody activity after inoculation with Ab2 2-17C3 were not protected against intracerebral challenge may be explained by the presence of only relatively low VN titers, which were obviously too low for effective protection. This was indicated by the protection experiment with Ab1 1-10C9E8, where an effective titer of ≥ 128 was shown to be necessary. It should be noted that the immunization method of these mice has not been optimized with regard to schedule, dose, and, perhaps more important, "antigenic" presentation. As has been shown to be the case with virus subunit vaccines, where monomeric forms of, for example, spike proteins of enveloped viruses were less immunogenic or even immunosuppressive in comparison with multimeric forms like micelles, virosomes, and, most recently, iscoms (MOREIN et al. 1984 and OSTERHAUS et al. 1985), it should be considered that also the presentation of Ab2 molecules or their idiotopes to the immune system in multimeric form might be more efficient. This was also independently suggested by the work of KENNEDY and DREESMAN (1984) in the HBV system and by the work of SHARPE et al. (1984) in the reovirus system. Experiments are presently being carried out in our laboratory in which multimeric forms, including iscom-like preparations of Ab2 2-17C3, prepared according to a new method developed by Karin LOVGREN and Bror MOREIN from the University of Uppsala (personal communication) are compared for their capacity to induce antipoliovirus neutralizing antibody and protection with noncomplexed soluble Ab2 2-17C3, as used in previous experiments.

Given the idea of JERNE (1984) that the immune system is "a hall of mirrors", i.e., self, anti-self, and subsequently anti-anti-self which reflects the outside antigenic universe, an immune response may be considered the reflection of events following a disturbance of the delicate dynamic balance of these components, by external or internal stimuli, that has to be compensated for by changing the complementary elements to a new state of equilibrium. In order to protect against e.g. a viral infection should one not feel more comfortable to have the immune system disturbed by a well-defined battery of its own elements, rather than by ill-defined conventional vaccines? However, for the near future we may not have a better choice than monoclonal mouse Ab2.

Acknowledgements. The authors are grateful to J. Urbain and J.-D. Franssen for stimulating discussions, and wish to acknowledge C. Kruyssen and T. de Haan for handling the manuscript.

References

Anonymous (1985) Outbreak of poliomyelitis in Finland. Commun Dis Reports 85/06:312
Arnon R, Shapira M, Jacob CO (1983) Synthetic Vaccines. J Immunol Methods 61:261
Beale AJ (1980) Cell substrate for killed poliovaccine production. Dev Biol Stand 47:19

Cazenave PA (1977) Idiotypic-anti-idiotypic regulation of antibody synthesis in rabbits. Proc Natl Acad Sci USA 74:5122
Cazenave PA, Roland J (1984) Internal images of rabbit immunoglobulin allotopes. Immunol Rev 79:139
Claflin JL, Davie JM (1974) Clonal nature of the immune response to phosphoryl choline IV. Idiotypic uniformity of binding site-associated antigenic determinants among mouse anti-phosphoryl choline antibodies. J Exp Med 140:673
Coutinho A (1980) The self non-self discrimination and the nature and acquisition of the antibody repertoire. Ann Immunol (Paris) 131D:235
Coutinho A, Forni L, Holmberg D, Ivars F (1983) Is the network theory tautologic? In: Moller G, Möller E (eds) Genetics of the immune response. Nobel sympsosium 55. Plenum, New York
Coutinho A, Forni L, Holmberg D, Ivars F, Vaz N (1984) From an antigen-centered, clonal perspective of immune responses to an organism-centered, network perspective of autonomous activity in a self-referential immune system. Immunol Rev 79:151
Emini EA, Jameson BA, Lewis AJ, Larsen GR, Wimmer E (1982) Poliovirus neutralization epitopes: analysis and localization with monoclonal antibodies. J Virol 43:997
Ertl HCJ, Finberg RW (1984) Sendai virus-specific T-cell clones: induction of cytolytic T cells by an anti-idiotypic antibody directed against a helper T-cell clone. Proc Natl Acad Sci USA 81:2850
Evans DM, Minor PD, Schild GS, Almond JW (1983) Critical role of an eight amino acid sequence of VP1 in neutralization of poliovirus type III. Nature 304:459
Fergusson M, Minor PD, Margrath DI, Qui Yi-Hua, Spitz M, Schild GC (1984) Neutralizing epitopes on poliovirus type 3 particles: an analysis using monoclonal antibodies. J Gen Virol 65:197
Francotte M, Urbain J (1984) Induction of anti-tobacco mosaic virus antibodies in mice by rabbit antiidiotypic antibodies. J Exp Med 160:1485
Franssen JD, Urbain J (1985) Some facts and thoughts about internal images. Int Rev Immunol (to be published)
Horodniceanu F, Crainic R, Barme M (1980) Cell substrate and risk in killed poliomyelitis vaccine. Dev Biol Stand 47:35
Jerne NK (1974) Towards a network theory of the immune system. Ann Immunol (Paris) 125C:373
Jerne NK (1976) The immune system: a web of V domains. Harvey Lect 70:93
Jerne NK (1981) In: Westen-Schnurr I (ed) Idiotypes-antigens on the inside. Editiones Roche, Basel, pp 12
Jerne NK (1984) Idiotypic networks and other preconceived ideas. Immunol Rev 79:5
Jerne NK, Roland J, Caznave PA (1982) Recurrent idiotopes and internal images. EMBO J 1:243
Ju ST, Benacerraf B, Dorf ME (1978) Idiotypic analysis of antibodies to poly (Glu 60 Ala 30 Tyr 10). Interstrain and interspecies idiotypic crossreactions. Proc Natl Acad Sci USA 75:6192
Ju ST, Cramer DV, Dorf ME (1979) Idiotypic analysis of anti-GAT antibodies V. Distribution of an interspecies cross-reactive idiotype. J Immunol 123:877
Kauffmann RS, Noseworthy JH, Nepom JT, Finberg R, Fields BN, Greene MI (1983) Syngeneic monoclonal anti-idiotypic antibody identifies the cell surface receptor for reovirus. J Immunol 131:2539
Kennedy RC, Dreesman GR (1983) Common idiotypic determinant associated with human antibodies to hepatitis B surface antigen. J Immunol 130:385
Kennedy RC, Dreesman GR (1984) Enhancement of the immune response to hepatitis B surface antigen. In vivo administration of antiidiotype induces anti-HBs that expresses a similar idiotype. J Exp Med 159:655
Kennedy RC, Sanchez Y, Ionescu-Matiu I, Melnick JL, Dreesman GR (1982) A common human anti-hepatitis B surface antigen idiotype is associated with the group A confirmation-dependent antigenic determinant. Virology 122:219
Kennedy RC, Dreesman GR, Sparrow JT, Culwell AR, Sanchez Y, Ionescu-Matiu I, Hollinger FB, Melnick JL (1983a) Inhibition of a common human anti-hepatitis B surface antigen idiotype by a cyclic synthetic peptide. J Virol 46:653
Kennedy RC, Ionescu-Matiu I, Sanchez Y, Dreesman GR (1983b) Detection of interspecies idiotypic cross-reactions associated with antibodies to hepatitis B surface antigen. Eur J Immunol 13:232
Leo O, Slaoui M, Marvel J, Milner ECB, Hiernaux J, Moser M, Capra JD, Urbain J (1984) Idiotypic analysis of polyclonal and monoclonal anti-p-A 20 phenylarsonate antibodies of BALB/c mice expressing the major crossreactive idiotype of the A/J strain. J Immunol 134:1734

Lindemann J (1973) Speculation on idiotypes and homobodies. Ann Immunol (Paris) 124C:171
Lindemann J (1978) Homobodies: do they exist? Ann Immunol (Paris) 130C:311
Lucas A, Henry C (1982) Expression of the major crossreactive idiotype in a primary anti-azobenzearsonate response. J Immunol 128:802
Meek K, Jeske D, Slaoui M, Leo O, Urbain J, Capra JD (1984) Complete amino acid sequence of heavy chain variable regions derived from two monoclonal anti-p-azophenylarsonate antibodies of BALB/c mice expressing the major cross-reactive idiotype of the A/J strain. J Exp Med 160:1070
Minor PD, Schild GC, Bootman J, Evans PMA, Fergusson M, Reeve P, Spitz M, Stanway G, Cann AJ, Hauptmann R, Clarke LD, Mountford RC, Almond JW (1983) Location and primary structure of a major antigenic site for poliovirus neutralization. Nature 301:674
Morein B, Sundquist B, Höglund S, Dalsgaard K, Osterhaus A (1984) Iscom, a novel structure for antigenic presentation of membrane proteins from enveloped viruses. Nature 308:457
Moser M, Leo O, Hiernaux J, Urbain J (1983) Idiotypic manipulation in mice: BALB/c mice can express the crossreactive idiotype of A/J mice. Proc Natl Acad Sci USA 80:4474
Nisonoff A, Lamoyi E (1981) Implications of the presence of an internal image of the antigen in anti-idiotypic antibodies: possible application to vaccine production. Clin Immunol Immunopathol 21:397
Norrby E (1983) Viral Vaccines: the use of currently available products and future developments. Arch Virol 76:163
Noseworthy JH, Fields BN, Dichter MS, Sobotka C, Pizer E, Perry LL, Nepom JT, Greene MI (1983) Cell receptors for the mammalian reovirus. I. Syngeneic monoclonal anti-idiotypic antibody identifies the cell surface receptor for reoviruses. J Immunol 131:2533
Osterhaus ADME, UytdeHaag FGCM (1985) Lymphocyte hybridomas: production and use of monoclonal antibodies. In: Spier R, Grissfiths B (eds) Animal cell biotechnology. Academic, New York (to be published) Proceedings of the workshop on abnormal cells, new products and risks (1984)
Osterhaus ADME, van Wezel AL, van de Marel P, van Steenis G (1981a) Production of monoclonal antibodies against poliomyelitis and rabies viruses. In: Peeters H (eds) Prot of the Biol F1. Proceedings of the twenty-ninth colloqium, vol 29, p 743, Pergamon Press, New York
Osterhaus ADME, Van Wezel AL, Van Steenis B, Drost GA, Hazendonk TG (1981b) Monoclonal antibodies to polioviruses: production of specific monoclonal antibodies to the sabin vaccine strains. Intervirology 60:218
Osterhaus ADME, Van Wezel AL, Hazendonk TG, UytdeHaag FGCM, Van Asten JAAM, Van Steenis B (1983) Monoclonal antibodies to polioviruses. Comparison of intratypic strain differentiation of poliovirus type I using monoclonal antibodies versus cross-absorbed antisera. Intervirology 20:129
Osterhaus A, Weijer K, UytdeHaag F, Jarrett O, Sundquist B, Morein B (1985) Induction of protective immune response in cats by vaccination with feline leukemia virus iscoms. J Immunol 135:1
Rajewski K, Takemori T (1983) Genetics, expression, and function of idiotypes. Annu Rev immunol 1:569
Reagan KJ, Wunner WH, Wiktor TJ, Koprowski H (1983) Anti-idiotypic antibodies induce neutralizing antibodies to rabies virus glycoprotein. J Virol 48,3:660
Roux KH, Metzger DW, Kazdin DS, Horng WJ (1984) Induced latent allotypes within rabbit anti-cross-reactive idiotype reagents. Direct immunoelectron microscopic evidence. Eur J Immunol 14:910
Sacks DL, Sher A (1983) Evidence that anti-idiotype induced immunity to experimental african trypanosomiasis is genetically restricted and requires recognition of combining site-related idiotopes. J Immunol 131,3:1511
Sacks DL, Esser KM, Sher A (1981) Immunization of mice against African trypanosomiasis using anti-idiotypic antibodies. J Exp Med 155:1108
Sacks DL, Kelsoe GH, Sachs DH (1983) Induction of immune responses with anti-idiotypic antibodies: implications for the induction of protective immunity. Springer Semin Immunopathol 6:79
Sharpe AH, Gaulton GN, McDade KK, Fields BN, Greene MI (1984) Syngeneic monoclonal anti-idiotype can induce cellular immunity to reovirus. J Exp Med 160:1195
Sinal NH (1982) Regulation of azophenylarsonate-specific repertoire expression. J Exp Med 156:1352

Sommé G, Roth C, Bismuth G, Leclercq L, Mazié JC, Théze J (1984) Public idiotypes and/or internal images in the anti-poly (Glu^{60} Ala^{30} Tyr^{10}) response: analysis by monoclonal antibodies. Ann Immunol (Paris) 135C:57

Strosberg AD (1983) Anti-idiotype and anti-hormone receptor antibodies. Springer Semin Immunopathol 6:67

Theze J, Moreau JL (1978) Genetic control of the immune response to GAT terpolymer I. Intrastrain and interspecies crossreactive idiotype. Ann Immunol (Paris) 129C:721

Urbain J, Wikler M, Franssen JD, Collignon C (1977) Idiotypic regulation of the immune system by the induction of antibodies against antiidiotypic antibodies. Proc Natl Acad Sci USA 74:5126

UytdeHaag FGCM, Osterhaus ADME (1985) Induction of neutralizing antiboy in mice against poliovirus type II with monoclonal antiidiotypic antibody. J Immunol 134:1225

UytdeHaag FGCM, Loggen HG, Logtenberg T, Lichtveld RA, Van Steenis B, Osterhaus ADME (1985a) Human peripheral blood lymphocytes from recently vaccinated individuals produce both homotypic and heterotypic neutralizing antibody upon in vitro stimulation with one type of poliovirus. J Immunol (to be published)

UytdeHaag F, Bunschoten H, Osterhaus A (1985b) Igh-V^a linked expression of a major cross reactive idiotope associated with reactivity against rabiesvirus glycoprotein: analysis by syngeneic monoclonal anti-idiotypic antibody. (to be published)

von Seefried A, Chun JH (1980) Serially subcultivated cells as substrates for poliovirus production for vaccine. Dev Biol Stand 47:25

Zoler ML (1984) Anti-idiotypic antibodies: a safer way to immunize against AIDS?. Biotechn 110:923

Molecular Mimicry of Parasite Antigens Using Anti-Idiotypic Antibodies

D.L. SACKS

1 Introduction: The Distinction Between Ab2α and Ab2β

Current interest in the molecular mimicry of microbial, transplantation, and tumor antigens using anti-idiotypic (anti-Id) antibodies is theoretically grounded in JERNE's network model (1974), which argues the following: The antigen-combining site, designated paratope, of every antibody molecule recognizes the idiotope of some other antibody molecule. This requires that external (non-immunoglobulin) epitopes and internal idiotopes must be largely overlapping, since they are each recognized by the same set of paratopes. Therefore, for each foreign epitope there must be an idiotope which bears its internal image. In order to obtain antibody molecules which bear epitope-related conformational structures, the strategy has been to immunize with antibody molecules, designated Ab1, expressing a paratope for a given epitope, with the intent of inducing a population of antibodies, designated Ab2, some of which share with the epitope a complementary structure to the Ab1 paratope (Fig. 1). This subpopulation of paratope-induced, anti-Ab1 antibodies has been termed Ab2β, to distinguish it from the more classically considered population of anti-Id antibodies, termed Ab2α, which are induced by Ab1 idiotopes and therefore lack any epitope-related structures (JERNE et al. 1982). Assuming that the interaction with receptors through either their paratopes or idiotopes can provide identical signals to the cells which bear them, then both Ab2 populations should be induced by Ab1. But because the number of possible bystander or framework-associated Ab1 idiotopes greatly exceeds the number of Ab1 paratopes, which are restricted by their common requirement for epitope complementarity, then

Laboratory of Parasitic Diseases, National Institute of Allergy and Infectious Diseases, National Institutes of Health, Bethesda, MD 20205, USA

Current Topics in Microbiology and Immunology, Vol. 119

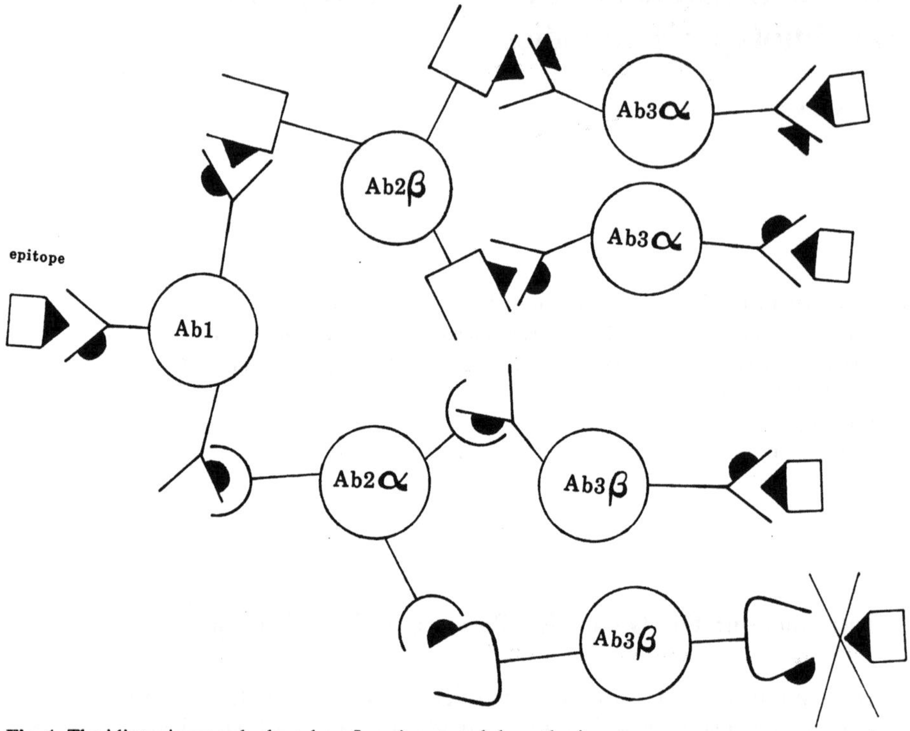

Fig. 1. The idiotypic cascade, based on Jerne's network hypothesis

paratope-induced Ab2β will generally represent a minor component of the Ab2 response.

To the extent that immunizations with anti-Id antibodies, more appropriately designated Ab2, have been successful in inducing specific responses to a wide spectrum of foreign antigens, including parasites, bacteria, viruses, alloantigens, and tumors, it is clear that operationally at least, Ab2 antibodies can behave as effective substitutes for antigen. However, whether or not these properties of Ab2 constitute molecular mimicry in the context which has just been described (i.e., shared conformational structures between Ab2β and the foreign epitope) remains speculative for virtually all of these systems. This is because Ab2α, despite their lack of epitope-related structures, can also activate antigen-specific clones via interaction with Ab1 idiotopes. The distinction between these two mechanisms will be discussed below.

The outcome of immunization with polyclonal Ab2, comprising both Ab2α and Ab2β, is depicted in Fig. 1. Consideration of the resulting Ab3 populations is restricted to those molecules which bear serologically definable Ab1 idiotopes and/or paratopes (often designated Ab1). This is done not just as a matter of convenience, but because the major outcome of immunization with polyclonal Ab2 is not to induce Ab3 antibodies which are anti-anti-Id, but rather molecules which bear at least some of the original Ab1 idiotopes, whether or not they also bind antigen. In other words, the response to polyclonal Ab2 (if we consider primarily the response to Ab2α) will be dominated by paratope-induced molecules (Ab3β) rather than idiotope-induced molecules of Ab2α. The reasons behind this apparent asymmetry have been discussed by a number of authors (Sacks et al. 1983; Paul and Bona 1982). We have suggested that because affinity-purified Ab2, which has generally been used for immunization, has been selected to be paratopically restricted to Ab1 idiotopes but idiotypically diverse, the resulting Ab3 will be dominated by antibody molecules complementary to Ab2 paratopes. In this way, Ab3, like Ab1, is selected to be complementary to the paratope of Ab2 merely by responding to the most frequent antigenic determinant, the paratope.

The induction of Ab3β expressing Ab1 idiotopes can be expected to include some molecules which also bind antigen, since in an autologous environment at least, clones possessing both properties must exist, and would be at the genetic level indistinguishable from the original Ab1. Induction of these molecules by Ab2α in syngeneic, allogeneic, or, in expectedly rare cases, xenogeneic environments would suggest the presence of recurrent Ab1 idiotopes which use the same or closely related germline genes. Ab3β should also comprise populations which bear some Ab1 idiotopes but do not bind antigen. This is so because whereas antigen-binding specificity is brought about by certain VDJ and VJ combinations, many individual idiotopes found on a given antibody need not be exclusively determined by these same gene combinations. In fact, Ab3 responses have often been shown to be dominated by idiotope-bearing but nonantigen-binding molecules. Subsequent immunization with antigen has in these cases been required to reveal that idiotope-bearing, antigen-specific clones have also been primed by Ab2 immunization (Bluestone et al. 1981; Bona et al. 1981; Cazenave 1977; Urbain et al. 1977).

In contrast to the response induced by Ab2α, the Ab3α response induced by epitope-related idiotopes of Ab2β should for the most part resemble Ab1 in terms of antigen binding. Because Ab2β behave like antigen they should not discriminate between different Ab1 molecules which each bind to the epitope but which might vary with respect to their idiotopes. Therefore, it is not possible in the absence of sequence data to determine whether or not antibodies induced by Ab2β use the same genes as the original Ab1.

In summary, there are two general mechanisms that need to be considered whenever Ab2 can be shown to activate antigen-specific clones: (a) the induction of recurrent Ab1 idiotopes by Ab2α and (b) the induction of Ab1 paratopes by Ab2β, which constitutes genuine molecular mimicry for which no genetic conclusions can be drawn. How can these possibilities be distinguished? The question is a critical one in the context of vaccine development, because it bears on the outcome of Id manipulations in outbred populations. For these purposes, it seems clear that Ab2β would represent the most useful form of Ab2, because their effects should not in most cases be restricted by genetic polymorphisms. The question will be addressed below in the context of Id manipulations in two parasite systems for which the induction of specific antibody has been demonstrated.

2 Induction of Immunity to African Trypanosomiasis: The Role of Ab2β

We have been able to induce protective immunity in mice against African trypanosomiasis (sleeping sickness) using anti-Id antibodies (SACKS et al. 1982; SACKS and SHER 1983). We have chosen experimental African trypanosomiasis as a model for Ab2-induced antimicrobial immunity largely because protective monoclonal antibodies defining a target Id can be easily raised against the variable surface antigens of these parasites. Allogeneic Ab2 were raised against two protective monoclonal Ab1, each with specificity for the major surface glycoprotein of a clone of *Trypanosoma rhodesiense,* WRATat 1.1. BALB/c mice were treated 4–5 weeks prior to infection with 20 μg affinity-purified Ab2. Table 1 shows that only treatment with anti-7H11 Ab2 was effective in immunizing mice against homologous challenge. There was no protection against infection with a parasite clone bearing another variant surface antigen.

Antisera made in BALB/c mice in response to infection or Ab2 treatment before infection were tested in a competitive radioimmunoassay for expression of each Id prior to a during the course of infection. Both the 7H11 and 11D5 idiotypes were produced by all BALB/c mice in response to infection. Thus, under normal unmanipulated conditions these represent public, cross-reactive, or recurrent Id, the expression of which has been demonstrated by classical genetic studies to be dependent upon genes linked to the heavy-chain allotype linkage group (SACHS 1981). In addition, both Id were readily inducible by their respective Ab2, since Ab1 idiotopes were detected in Ab2 treated mice prior to exposure to antigen. Because the concentrations of induced Id were relatively low, it was not possible to determine directly what proportion had antigen-binding activity, since assays for antibody were far less sensitive than those for Id detection. However, the immunity induced by 7H11 Ab2 treatment was associated with the more rapid and enhanced expression of the 7H11 Id by 4 days after infection. Thus, the immunity appears to have been largely due to the priming of Id-positive antigen-specific clones. In contrast, immunization of BALB/c mice with Ab2 specific for 11D5 appears to have had relatively

Table 1. Ab2 immunization of BALB/c Mice againt WRATat 1.1 challenge

Expt	Ab2 administered		
	NMIg	7H11	11D5
1	0/10[a]	8/10	2/10
2A	0/9	6/9	0/8
2B[a]	0/8	0/8	NT

Number of mice without VAT 1.1 in primary parasitemia/number tested
NT, Not tested
[a] Challenge with 100 NIHT at 1 trypanosomes

little effect on the priming of Id-bearing antigen-specific clones, since these mice were not immune and the expression of this Id in treated mice was not significantly increased after challenge. This data suggests that the relative proportion of Id-bearing clones activated that also had antigen-binding activity differed significantly between the two Ab2. One explanation for this result could be that 7H11 Ab2 contained a relatively greater concentration of Ab2*β* than did 11D5 Ab2. While it is not possible to formally quantitate the concentration of Ab2*β* within a polyclonal Ab2 population, it is possible to determine the concentration of the total population of combining site related Ab2. A minimal requirement for Ab2*β* is that it should compete with antigen for binding to Ab1, although is is clear that because of steric hindrance, molecules which fulfill this property are not necessarily Ab2*β*. Binding studies is which the respective Ab2 were used to compete with purified variant surface antigen for binding to the homologous monoclonal Ab1 comfirmed that 7H11 Ab2 had concentrations of combining-site related antibodies more than six fold higher than the 11D5 Ab2. Thus, because relatively more of the idiotopes defined by 7H11 Ab2 appear to be closer to or within the combining site, a larger proportion of clones activated which bear these idiotopes might also be expected to be antigen binding. In studies by Takemori et al. (1982) in the NP system, similar results were obtained after immunization with cross-linked monoclonal Ab2. Immunization with a hapten-inhibitable Ab2 induced Id-bearing molecules, essentially all of which were NP binding. In contrast, immunization with a monoclonal Ab2 defining an idiotope not intimately associated with NP-binding site resulted in the induction of high levels of idiotope-bearing molecules, only a small fraction of which were NP binding.

If Ab2*β* are indeed responsible for the induction of immunity in the trypanosome system, then these effects should not be genetically restricted, since Ab2*β* should not discriminate between different Ab1 with specificity for the same epitope-related structure. Genetic control of Id induction was investigated in the mouse strains listed in Table 2, each receiving 20 μg 7H11 Ab2 6–8 weeks prior to challenge with parasite clone WRATat 1.1. Only mice that bore the a-heavy-chain allotype demonstrated immunity. Allotype congeneic recombinant mice (CB20) had infections indistinguishable from those of nonimmunized BALB/c mice. Antisera made in these inbred mouse strains in response to infection or Ab2 treatment were tested for expression of the 7H11 Id. This Id was produced only by BALB/c and BALB. B mice in response to either infection or Ab2 immunization. The simplest interpretation of the data would imply polymorphism of the structural genes encoding the Id phenotype. Similar genetic control over Ab2-induced Id expression has been demonstrated in other systems (Trenkner and Riblet 1975; Bluestone et al. 1981; Epstein et al. 1982; Takemori et al. 1982). The results argue against the presence of Ab2*β*, since the primary effect of variable regions (V) polymorphism should be on the activities of idiotope-restricted Ab2*α*. Clearly, no amount of Id selection via anti-Id will be successful unless the appropriate information encoding such V sequences are present in the genome. How can this data be reconciled with the results of the binding studies which argued for the existance of Ab2*β*? It is of course possible that the related epitope borne by Ab2*β* is not recognized by other

Table 2. Ab2 immunization of various mouse strains against WRATat 1.1 challenge

Mouse strain[a]	Igh-C	H-2	Immune/total[b]
BALB/c	a	d	8/10
BALB.B	a	b	4/8
C57BL/10	b	b	0/10
A/J	e	a	0/10
C.B20	b	d	0/10

[a] All mice immunized with 20 μg 7H11 Ab2 6 weeks before challenge with 100 WRATat 1.1

[b] Immunity determined as the absence of VAT 1.1-bearing parasites in primary parasitemia

mouse strains, either on Ab2β or on the parasite antigen itself. This question remains unresolved.

3 Further Evidence for Ab2β: Molecular Mimicry of a Carbohydrate Epitope of *Trypanosoma Cruzi*

NISONOFF and LAMOYI (1981) have suggested that stronger proof of the existence of Ab2β would be forthcoming if upon inoculation into phylogenetically distant species, Ab2 could induce specific antibodies. This has in fact been demonstrated by URBAIN et al. (1984), who were able to induce antitobacco mosaic virus (TMV) antibodies in mice using Ab2 prepared against a rabbit Id. Similarly, KENNEDY et al. (1984) have induced antibodies in mice against hepatitis B surface antigen (HBsAg) using Ab2 prepared against a common Id shared by human antibodies to HBsAg. While we have not pursued such experiments in the African trypanosome system, we have been able to achieve similar results in another parasite model, *Trypanosoma cruzi.* These protozoan parasites are the etiologic agents of Chagas' disease, a chronic debilitating disease in South and Central America. A glycoprotein with an apparent mol. w. of 72000 specific for the insect stages of *T. cruzi* has been identified (SNARY et al. 1981), which protects mice against insect-derived trypomastigote challenge (SNARY 1983) and also appears to be involved in the control of stage-specific morphogenesis (SHER and SNARY 1982). A monoclonal mouse antibody (29.26) has been produced which recognizes an unusual carbohydrate epitope on the 72K glycoprotein (SNARY et al. 1981). We have raised a xenogeneic Ab2 in rabbits with the intent of formally demonstrating that molecular mimicry of a carbohydrate epitope can be achieved. This may be of practical relevance, since many microbial antigens contain critical carbohydrate epitopes which cannot be easily reproduced by recombinant DNA technologies. STEIN and SODERSTROM (1984) have recently demonstrated that a monoclonal Ab2 could prime weanling mice for antipolysaccharide responses and protect against infection with pathogenic *E. coli.*

Table 3. Induction of antibodies to *T. cruzi* in different species using 29.26 Ab2

Species	Immunogen	Reciprocal IFA titer	
		Tulahen strain	Brazil strain
Mouse (BALB/c)	29.26 Ab2	640[a]	<10
	Normal Rab IgG	<10[a]	<10
Rabbit	29.26 Ab2[b]	80,40	<10
	Normal Rab IgG[b]	<10	<10
Guinea pig	29.26 Ab2[c]	160[d]	NT
	Normal Rab IgG	<10[d]	NT

[a] Geometric mean titer of eight mice
[b] Rabbits immunized with 500 μg 29.26 Ab2 or NRIgG coupled to KLH in CFA
[c] Guinea pigs immunized with 300 μg 29.26 Ab2 or NRIgG in CFA
[d] Geometric mean titer of four guinea pigs

In the *T. cruzi* system, immunization of BALB/c mice with affinity-purified rabbit Ab2 in complete Freund's adjuvant (CFA) induced high titers of antibodies against the 72K glycoprotein. Immunoprecipitation of ^{125}I-surface labeled parasites by antisera from Ab2 immunized mice revealed the same pattern on SDS-polyacrylamide gels as did immunoprecipitation using 29.26. When assayed on Tulahen strain *T. cruzi* epimastigotes, the geometric mean indirect immunofluorescent antibody titer (IFA) of BALB/c mice immunized with 29.26 Ab2 was 1/640 (Table 3). When assayed on a strain of *T. cruzi* (Brazil), which contains the 72K glycoprotein, but lacks exposed 29.26 epitope (Kirchhoff et al. 1984), the antisera were negative, suggesting that the antibody induced by Ab2 was indeed specific for a carbohydrate determinant. The 29.26 Id was found to be a recurrent mouse Id because it was expressed by all BALB/c mice in response to infection as well as Ab2 treatment (Table 4). It is probable that the high levels of both the 29.26 Id and anti-72K specific antibodies induced by Ab2 treatment in BALB/c mice did reflect at least in part the activities of conventional Ab2α recognizing and inducing clones bearing the recurrent 29.26 idiotype. However, the induction of specific antibodies, albeit of lower titers, in both rabbits and guinea pigs (Table 3) following immunization with Ab2 suggests either the 29.26 Id represents an interspecies cross-reactive Id, or the presence of Ab2β within the Ab2 population. As suggested by Nisonoff and Lamoyi (1981), support for the latter possibility would be strengthened still further if the Ab2 can be shown to react with conventional anti-72K antibodies from various sources. In this case the reactivity would not be due to Id selection, but antigen selection. In Table 4, anti-72K antibodies from rabbits immunized with affinity-purified 72K glycoprotein or humans infected with *T. cruzi* and known to contain anti-72K antibodies based on immunoprecipitation of surface-labeled parasites, were used to inhibit the binding of ^{125}I-labeled 29.26 to solid phase Ab2. In each case, significant inhibition of binding was observed compared with controls. Thus, the Ab2 in these assays behaved as though it contained antigen.

Table 4. Levels of 29.26 idiotype in anti- *T. cruzi* antisera from different species

Species	Imunogen	29.26 Idiotype (μg/ml)[a]
BALB/c Mice	Preimmune	<0.1[b]
	T. cruzi infection[c]	1.2[b]
	29.26 Ab2	6.5[b]
	29.26 Ab2 followed by *T. cruzi* infection	27.0[b]
Rabbit	Control	<0.1, <0.1
	Affinity-purified 72K	2.2, 0.7
Human	Control	<0.1, <0.1, <0.1
	T. cruzi infection	0.8, 1.1, 0.8

[a] Idiotype levels determined by the ability of the antisera to inhibit the binding of ^{125}I-labeled 29.26 to affinity purified Ab2. Assay was standardized using known concentrations of cold 29.26 as inhibitor

[b] Arithmetic mean value of eight mice per group

[c] Infection with 10^5 Tulahen strain metacylic trypomastigotes

Similar results have been reported by URBAIN et al. (1984) in which Ab2 recognized anti-TMV antibodies from all rabbits, chickens and mice; KENNEDY et al. (1983) in which Ab2 recognized anti-HBsAg antibodies from six other species; UYTDEHAAG and OSTERHAUS (1985) in which monoclonal Ab2 recognized antipolio virus type II antibodies of rats, guinea pigs, mice, and humans; and STEIN and SODERSTROM (1984) in which monoclonal Ab2 recognized anti-*E. coli* polysaccharide antibodies of both mouse and rats. Defining Ab2β by this criteria may in fact be more meaningful than the induction of antibody by Ab2 in different species, since, as has been recently reported by FRANCOTTE and URBAIN (1984), Id can be induced across a species barrier using Ab2 that do not recognize conventional antibodies from different species, and therefore do not display the properties of Ab2β which have just been described. In this case rabbit Ab2 raised against a private rabbit TMV Id when coupled with lipopolysaccharide (LPS) were able to induce anti-TMV antibodies in BALB/c mice. Thus, even though the Ab2 does not appear to contain Ab2β, it is able to circumvent the genetic restrictions normally associated with Ab2α because it seems able to recognize and activate normally "silent" interspecies cross-reactive idiotopes. Activation of "silent" idiotypes within syngeneic or allogeneic systems has also been described (CAZENAVE 1977; URBAIN et al. 1977; BONA et al. 1981). These data suggest that the absence of expression of a particular Id, even when different species are compared, may not necessarily reflect the absence of requisite structural genes, but may indicate that Id expression is subject to regulation. Administration of Ab2, especially in conjunction with LPS, appears to bypass these regulatory processes. If "silent" interspecies cross-reactive idiotopes are indeed a more general feature of the potential immune repertoire, then their induction by Ab2 in outbred populations would no longer be dependent upon the presence of Ab2β. However, there are a number of studies, mentioned previously, in which the outcome of Id manipulations were found to be strictly strain specific (TRENKNER and RIBLET 1975; TAKEMORI et al.

1982; BLUESTONE et al. 1981). And as has also been mentioned, the induction of immunity to African trypanosomiasis using Ab2 was genetically restricted. Of course it can be argued that in these studies the use of Ab2 to reveal "silent" clones was not carried out under appropriate conditions (i.e., the Ab2 were not coupled to immunogenic carriers or were not given in conjunction with mitogen stimulation). One further point regarding recurrent interspecies idiotopes needs to be mentioned. If Ab2 can select for the expression of interspecies idiotopes that are normally "silent," then the appearance in conventional antisera from different sources of antibodies which react with Ab2 might rather reflect the presence of interspecies cross-reactive idiotopes normally expressed, than an effect of Ab2β. Clearly, these two possibilities will only be distinguished when Ab2-induced antibodies from various species are affinity purified and compared with the original Ab1 in terms of sequence or isoelectric focusing (IEF) patterns.

4 Summary

The elicitation of specific antibodies using Ab2 instead of antigen has now been demonstrated for responses to most classes of microbial antigens. We have been able to induce specific antibody against two different pathogenic protozoa, African trypanosomes and *T. cruzi,* using polyclonal Ab2. It has been far more difficult to ascribe these phenomena to molecular mimicry, which is an inherent property of only a subpopulation of Ab2, designated Ab2β. The issue is not a trivial one, for it pertains to the outcome of idiotypic manipulations in outbred populations – clearly a major consideration in the context of vaccine development. Because paratope-induced Ab2β by definition bear a related epitope, they should not discriminate between different Ab1 and their effects should not be generally restricted by genetic polymorphisms. The induction of immunity to African trypanosomiasis in mice was, however, restricted by genes linked to the heavy-chain allotype locus, suggesting that Id-recognizing Ab2α were primarily if not exclusively responsible for these effects. The alternative possibility that those mouse strains which failed to respond to Ab2 treatment were unable to produce antibodies which could recognize either the epitope-related structure on Ab2β or on the antigen itself cannot be excluded. Evidence for an effect of Ab2β can be strengthened if the administration of Ab2 can be shown to elicit the production of antigen-binding molecules in different species, and if Ab2 can be shown to react with conventional antisera from different sources. Both properties of Ab2 were demonstrated in another parasite model, in which rabbit Ab2 were produced against a recurrent BALB/c Id with specificity for a carbohydrate epitope of a surface glycoprotein of *T. cruzi.* The Ab2 induced specific antibodies in mice, rabbits, and guinea pigs, and reacted with anti-*T. cruzi* antibodies produced in rabbits as well as humans. The behavior of this Ab2 is therefore consistent with an internal-image hypothesis, and suggests that molecular mimicry of carbohydrate epitopes can be easily achieved. Since the existence of an interspecies cross-reactive Id cannot formally be ruled out, more rigorous proof of molecular mimicry in these systems will require sequencing of affinity-purified Ab2-induced antibodies.

References

Bluestone JA, Epstein SL, Ozato K, Sharrow SO, Sachs DH (1981) Anti-idiotypes to monoclonal anti-H-2 antibodies. II. Expression of anti-H-$2K^k$ idiotypes on antibodies induced by anti-idiotype or H-$2K^k$ antigen. J Exp Med 154:1305

Bona C, Heber-Katz E, Paul WE (1981) Idiotype-anti-idiotype regulation. I. Immunization with a levan-binding myeloma protein leads to the appearance of auto-anti antibodies and to the activation of silent clones. J Exp Med 153:951

Cazenave PA (1977) Idiotypic-anti-idiotypic regulation of antibody synthesis in rabbits. Proc Natl Acad Sci USA 77:5122

Epstein SL, Masakowski VR, Sharrow SO, Bluestone JA, Ozato K, Sachs DH (1982) Idiotypes of anti-IA antibodies. II. Effects of *in vivo* treatment with xenogeneic anti-idiotype. J Immunol 129:1545

Francotte M, Urbain J (1984) Induction of anti-TMV antibodies in mice by rabbit anti-idiotype antibodies. J Exp Med 160:1485

Jerne NK (1974) Towards a network theory of the immune system. Ann Immunol (Paris) 125c:373

Jerne NK, Roland J, Caznave PA (1982) Recurrent idiotypes and Internal images. E M B O 1:234

Kennedy RC, Ionescu-Matiu RC, Sanchez Y, Dreesman GR (1983a) Detection of interspecies cross-reactions associated with antibodies to Hepatitis B surface antigen. Eur J Immunol 13:232

Kennedy RC, Adler-Strothz K, Henkel RD, Sanchez Y, Melnick JL, Dreesman GR (1983b) Immune response to hepatitis B surface antigen: and enhancement by prior injection of antibodies to the idiotype. Science 221:835

Kirchoff LV, Hieny S, Shriver GM, Snary D, Sher A (1984) Cryptic epitope explains the failure of monoclonal antibody to bind to certain isolates of *Trypanosoma cruzi*. J Immunol 133:2731

Nisonoff A, Lamoyi E (1981) Implications of the presence of an internal image of the antigen in anti-idiotypic antibodies; possible application to vaccine production. Clin Immunol Immunopathol 21:397

Paul WE, Bona C (1982) Regulatory idiotopes and immune networks: a hypothesis. Immunol Today 3:230

Reagan KJ, Wunner WH, Wiktor IJ, Koprowski H (1983) Anti-idiotypic antibodies induce neutralizing antibodies to rabies virus glycoprotein. J Virol 48:660

Sachs DH (1981) Genetic control of idiotype expression. In: Progress in immunology IV, J. Dausset, ed. Academic, London

Sacks DL, Sher A (1983) Evidence that anti-idiotype induced immunity to experimental African trypanosomiasis is genetically restricted and requires recognition of combining site-related idiotopes. J Immunol 131:1511

Sacks DL, Esser KM, Sher A (1982) Immunization of mice against African trypanosomiasis using anti-idiotypic antibodies. J Exp Med 155:1108

Sacks DL, Kelsoe GH, Sacks DH (1983) Induction of immune responses with anti-idiotypic antibodies: implications for the induction of protective immunity. Springer Semin Immunopathol 6:79

Sher A, Snary D (1982) Specific inhibition of the morphogenesis of *Trypanosoma cruzi* by a monoclonal antibody. Nature (Lond) 300:639

Snary D (1983) Cell surface glycoproteins of *Trypanosoma cruzi:* protective immunity in mice and antibody levels in human chagasic sera. Trans R Soc Trop Med Hyg 77:126

Snary D, Ferguson MAJ, Scott MT, Allan AK (1981) Cell surface antigens of *Trypanosoma cruzi:* use of monoclonal antibodies to identify and isolate an epimastigote specific glycoprotein. Mol Biochem Parasitol 3:343

Stein K, Soderstrom T (1984) Neonatal administration of idiotype or anti-idiotype primes for protection against *E. coli* K13 infection in mice. J Exp Med 160:1001

Takemori T, Tesch H, Reth M, Rajewski K (1982) The immune response against anti-idiotope antibodies I. Induction of idiotope-bearing antibodies and analysis of the idiotope repertoire. Eur J Immunol 12:1040

Trenkner E, Riblet R (1975) Induction of antiphosphorylcholine antibody formation by anti-idiotypic antibodies. J Exp Med 142:1121

Urbain J, Wikler M, Franssen JD, Collignon C (1977) Idiotypic regulation of the immune system by the induction of antibodies against anti-idiotypic antibodies. Proc Natl Acad Sci USA 74:5126

Urbain J, Slaoui M, Mariame B, Leo O (1984) Idiotypy and internal images. In: Idiotypy in biology and medicine, H. Köhler, J. Urbain, P.A. Cazenave, eds. Academic, London

UytdeHaag F, Osterhaus A (1985) Induction of neutralizing antibody in mice against poliovirus type II with monoclonal anti-idiotypic antibody. J Immunol 134:1225

Wikler M, Franssen JD, Collignon C, Leo O, Mariane B, Van de Walle P, De Groote D, Urbain J (1979) Idiotypic regulation of the immune system: common idiotypic specificities between idiotypes and antibodies raised against anti-idiotypic antibodies in rabbits. J Exp Med 150:184

Network Regulation of the Immune Response to Bacterial Polysaccharide Antigens

K.E. STEIN

1 Introduction

Protective immunity against encapsulated bacteria can be achieved by eliciting serum antibodies directed against the polysaccharide capsule (SCHNEERSON et al. 1971; KAIJSER and AHLSTEDT 1977; EGAN et al. 1983). In adults and in children over the age of 2 years this can readily be accomplished by the use of polysaccharide vaccines (MAKELA et al. 1977; PARKE et al. 1977; MAKELA et al. 1980; VODOPIJA et al. 1983). Infants, however, are unresponsive to most polysaccharide antigens (SMITH et al. 1973; PARKE et al. 1977; PELTOLA et al. 1977; PINCUS et al. 1982; KAYHTY et al. 1984), and a significant proportion of invasive and other serious bacterial infections caused by organisms such as *Haemophilus influenzae* type b (Hib; KAYHTY et al. 1984), *Neisseriae meningitidis* groups A, B, and C (BAND et al. 1983) and *Streptococcus pneumoniae* (MAKELA et al. 1980; DOUGLAS and MILES 1984) occurs in this age group. Stimulation of antipolysaccharide antibodies in infants would be enormously important both from the medical point of view and from the inferences that can be drawn regarding our understanding of the mechanisms of neonatal unresponsiveness.

Division of Bacterial Products, Office of Biologics Research and Review, Center for Drugs and Biologics, Food and Drug Administration, 8800 Rockville Pike, Bethesda, MD 20205, USA

Current Topics in Microbiology and Immunology. Vol. 119

In this chapter I will discuss some of the general features of the immune response to polysaccharide antigens which form the background for the development of experiments designed to regulate the antibody response to these antigens via network manipulation. I will then review the experiments done in my laboratory and those of my collaborators, Constantin Bona and Tommy Soderstrom, that involve neonatal priming with idiotypes (Id) and anti-Id and that have resulted in the induction of protective immunity against infection with *Escherichia coli* K13. I will conclude with a discussion of the possible mechanisms involved in neonatal priming, in view of what we know about other systems, and speculate on the future of this area of investigation. It is not my intention to completely review the field of immunity to polysaccharide antigens, but rather to review the salient features which suggest a need for studies involving network manipulation and to demonstrate that such an approach can successfully lead to the induction of protective immunity against bacterial infection. Anti-Id antibodies and the induction of protective immunity in viral and parasitic infections will be discussed in other chapters in this book.

2 Background Studies

2.1 Introduction

Immune responses to polysaccharide (PS) antigens have several characteristics that distinguish them from the antibody response to protein antigens. These include thymus independence (TI, MOSIER et al. 1977a), failure to stimulate a memory response and to undergo affinity maturation (BRILES and DAVIE 1975; KAYHTY et al. 1984), late development in ontogeny (MOSIER et al. 1977a; BAKER et al. 1977; PELTOLA et al. 1977; GOLD et al. 1977; BONA et al. 1978; BONA et al. 1979; KAYHTY et al. 1984), the requirement for a specific subset of B lymphokytes (MOSIER et al. 1977b), isotype restriction (YOUNT et al. 1968; PERLMUTTER et al. 1978; DER BALIAN et al. 1980), and Id restriction (KUNKEL et al. 1963; CRAMER and BRAUN 1975; PERLMUTTER et al. 1977; HANSBURG et al. 1979; STEIN et al. 1980; WIKLER and URBAIN 1984). Approaches to overcoming the problem of neonatal unresponsiveness to PS antigens have been in two general areas: the alteration of PS antigens to make them thymus dependent (TD), and specific triggering of anti-PS antibody-forming cells via the use of Id and anti-Id priming.

2.2 Thymus Dependence and B Cell Subset Restriction

In efforts to convert immune responses to PS antigens to thymus dependence (TD), both oligosaccharides (JORBECK et al. 1981; SVENSON and LINDBERG 1981; JENNINGS and LUGOWSKI 1981; STEIN et al. 1982; ANDERSON 1983; JENNINGS et al. 1984) and polysaccharides (SCHNEERSON et al. 1980; BEUVERY et al. 1983; GRANOFF et al. 1984) have been coupled to protein carriers. Such procedures have resulted in conjugates that stimulated anti-PS antibodies. *H. influenzae* type b conjugates stimulated anti-PS antibodies in mice that were poorly or not at all responsive to the PS alone (SCHNEERSON et al. 1980; CHU et al. 1983), in weanling rabbits (ANDERSON 1983), in infant monkeys (SCHNEERSON et al. 1984), and in human infants over the age of 6 months (ANDERSON and PICHICHERO 1984; WARD et al. 1984; ZAHRADNIK and GORDON 1984; LEPOW et al. 1985). Studies of the immunogenicity of Hib-protein conjugates in large numbers of infants below the age of 6 months have not yet been reported.

In the B512 dextran system, where the PS itself is immunogenic in adults, the response to an oligosaccharide-protein conjugate could be compared with that to the PS in mice of different ages (STEIN et al. 1983). It was shown that the conjugate was capable of stimulating antidextran antibodies in mice that were too young to respond to dextran, itself. The antibody response, however, was still dependent on the presence of the Lyb5$^+$ subset of lymphocytes (a subpopulation of cells originally described by AHMED et al. (1977) and HUBER et al. (1977)) as had been shown earlier for the response to PS alone (MOSIER et al. 1977a). As a result, the dextran oligosaccharide protein conjugate failed to stimulate antidextran antibodies in newborn mice, and the development of antidextran antibodies followed the development of Lyb5$^+$ cells. These studies suggested the need for another means of stimulating anti-PS antibodies in newborns, one which might be capable of stimulating the Lyb5$^-$ subset of lymphocytes which are present at birth. The Lyb5$^+$ subset develops at 7–10 days after birth in mice and reaches adult levels at 3–4 weeks (MOSIER et al. 1977b). The equivalent subset in man has not yet been described, but a recent report (GOLDING et al. 1984) suggests that the B cells from human newborns are functionally equivalent to Lyb5$^-$ cells in newborn mice.

The ability to stimulate B cells in newborns is important for two reasons. The first is to prevent Lyb5$^-$ cells from being tolerized.

When we described (STEIN et al. 1983) the restriction of anti-PS antibody forming cells to the Lyb5$^+$ subset, regardless of whether the antigen was given in a TI or TD form, we hypothesized that the Lyb5$^-$ subset did not produce anti-PS antibodies because those cells were tolerized by environmental PS antigens. The Lyb5$^+$ subset could respond to environmental PS antigens and the degree of expression of cells of any individual specificity would then depend on the environmental exposure, most likely through colonization of the gut. BJÖRKSTEN et al. (1982) studied the development of dextran-reactive antibodies in healthy infants and suggested that the presence of such antibodies is related to the intestinal flora. One means by which the tolerization of the Lyb5$^-$ subset might be prevented would be to prime these cells with anti-Id or Id, via anti-idiotypic helper T cells (reviewed by BOTTOMLY 1984), before exposure to the environmental PS. A second reason for attempting to prime B cells in newborns is that infants born to mothers with low levels of anti-PS antibodies would be protected from disease caused by encapsulated organisms in the absence of adequate maternal immunity, in the form of passively acquired IgG and IgA antibodies.

2.3 Idiotype Restriction

The antibody response to polysaccharide antigens is Id-restricted (or pauciclonal) resulting in only one or a few of the possible antibody clones being expressed following immunization. An extreme example of this is the anti-inulin (In) response to the $\beta(2\rightarrow1)$ fructosan branch determinant present on bacterial levan (BL). Following immunization with BL, virtually all BALB/c mice produce anti-In antibodies of a single isoelectrofocusing (IEF) spectrotype (STEIN et al. 1980) that is identical to the J606 In-binding myeloma protein (RUBINSTEIN et al. 1983). In general, the antibody response to a polysaccharide antigen is distinguished by characteristic Id as seen for BL (LIEBERMAN et al. 1979), group A streptococcal polysaccharide (BRILES and DAVIE 1975), and the $\alpha(1\rightarrow3)$ dextran determinant (HANSBURG et al. 1976). For consideration of network manipulation experiments, this fact is important in several ways. The first is that most of the response to a given PS antigen might be regulated by altering a single Id, although there are clearly other important factors. Regulatory genes, not linked to the major histocompatibility complex or the immunoglobulin locus, have been described in the response to In (STEIN et al. 1980), streptococcal group A carbohydrate (CRAMER and BRAUN 1975), and to the phosphorylcholine (PC) determinant of the C polysaccharide of R36A *S. pneumoniae* (CANCRO et al. 1978). These may play an important role in outbred populations and should be further investigated.

The second reason that Id restriction is important, from an experimental point of view, is that manipulation of a normally "silent" Id, such as the A48 Id in BALB/c mice immunized with BL (HIERNAUX et al. 1981), allows the response to be monitored on a relatively low background and affords a means of studying the mechanisms of Id regulation. Details of the A48 system will be covered below.

Most important in the consideration of network manipulation as a means of stimulating protective immunity to the capsular polysaccharides of pathogenic organisms, is the ability to regulate Id that are, themselves, protective. BRILES et al. have demonstrated that anti-PC antibodies of different Id of both the IgM (BRILES et al. 1982) and IgG (BRILES et al. 1984) isotypes differ in their abilities to confer passive immunity to *S. pneumoniae*. They found that for both isotypes, molecules bearing the T15 Id, which is the dominant Id in the BALB/c response to R36A *S. pneumoniae*, were more protective than molecules of the M603 or M511 Id. The fact that most anti-PS responses are Id restricted should facilitate the determination of the relevant protective Id for a given encapsulated organism.

3 Neonatal Priming for Antipolysaccharide Responses

3.1 Introduction

A means of priming newborns for an antibody response following exposure to PS antigens is clearly desirable. As discussed above, an area of active investigation is the use of oligo- or polysaccharide-protein conjugates to stimulate a TD response to PS. An alternative approach is the use of anti-Id antibodies. The following are four reasons for using anti-Id: (a) They are part of the normal regulatory pathway; (b) They are effective in situations where antigen is not, for example, in neonates and in the activation of "silent" Id; (c) They can be used in situations where purified antigen is not available or is harmful; (d) They are selective and can be used to turn on specific antibody clones.

The first point has been demonstrated in several systems. BONA et al. (1978) demonstrated that the plaque-forming cell (PFC) response to BL over a 60-day period following immunization occurred in two peaks, at 10 and 30 days. The peaks of E109, a dominant Id in the BL system, parallelled the anti-BL PFC peaks. They also demonstrated that there were anti-E109 PFC following immunization with BL and that there was an inverse corrrelation between the number of E109 positive PFC and the number of anti-E109 PFC. They concluded that the response to BL was regulated by auto-anti-Id antibodies. Subsequently, FERNANDEZ and MOLLER (1979 and 1980) demonstrated that the response to the PS, B512 dextran, was under auto-anti-Id regulation and SCHRATER et al. (1979) showed that the normal immune response to trinitrophenyl-Ficoll (a synthetic PS) was regulated by auto-anti-Id antibodies. Recently, RODKEY et al. (1983) have demonstrated auto-anti-Id regulation of the rabbit response to *Micrococcus lysodeikticus* and GEHA (1983) and SAXON and BARNETT (1984) have demonstrated auto-anti-Id regulation of the TD response to tetanus toxoid in humans.

Point (c) is particularly important as far as the suitability of anti-Id as vaccines is concerned. With regard to vaccines made from antigens present in pathogenic organisms, one has to take into account the lack of availability of large quantities of purified antigen. NISONOFF and LAMOYI (1981) pointed this out in their article on anti-Id as vaccines, for complex antigens from pathogens such as parasites, but it applies equally to bacterial capsular polysaccharides. In addition, bacterial toxins, even if purified in large quantities, have the problem of being harmful if not properly toxoided. In all of these cases anti-Id eliminates the necessity of having large quantities of purified antigen.

Finally, in seeking vaccines that are protective against disease, one can choose anti-Id directed against Id which are themselves protective, such as T15 which provides passive protection against *S. pneumoniae* (see BRILES et al. 1982, 1984). By using anti-Id, one may also have the opportunity to select the isotype as well, as has been suggested by the recent studies of NISHINARITA et al. (1985) which demonstrated the existence of an Id determinant created by the T15 IdX and a portion of the constant region of IgA. This Id was detected by a monoclonal anti-Id and the reaction was inhibited only by IgA $T15^{+}$ anti-PC antibodies and not by $T15^{+}$ antibodies of other isotypes. Thus, in biologically important systems where isotype may play an important role for functions such as complement fixation and opsonic activity, the use of appropriately selected anti-Id could result in the induction of not only the desired combining site but also the desired isotype.

3.2 Neonatal Priming for the "Silent" Id, A48

A48 is a BALB/c myeloma protein that binds BL (LIEBERMAN et al. 1975). The A48 Id is not normally expressed in BALB/c mice following immunization with BL but it can be induced in nu/nu BALB/c mice if they are pretreated

Table 1. Requirement of antigen for the activation of the A48 Id⁺ response in BALB/c mice treated at birth with anti-A48 Id antibody (from RUBINSTEIN et al. 1983, by copyright permission of The Rockefeller University press)

Mice treated at birth with	Mice challenged with	Anti-BL PFC/spleen[a]		Anti-TNP PFC/spleen[b]	
		Total response (*n*)	A48 Id⁺ (%)	Total response (*n*)	M460 Id⁺ (%)
Nil	20 μg BL	2,180 ± 513	6 ± 3	<100	ND
Nil	20 μg TNP-Ficoll	<100	ND	44,800 ± 3,582	16 ± 7
0.01 μg anti-A48 Id	Nil	<100	ND	<100	ND
0.01 μg anti-A48 Id	20 μg BL	2,920 ± 718	52 ± 7	<100	ND
0.01 ng anti-A48 Id	20 μg TNP-Ficoll	<100	ND	39,700 ± 7,860	24 ± 6

ND, not determined, insufficient number of plaques

[a] Total anti-BL PFC/spleen response was measured 5 days after immunization. The percentage A48 Id⁺ anti-BL PFC was determined by the addition of anti-A48 Id antisera to the agarose and scoring the difference in the number of PFC without antisera. Results represent the mean ± SEM for the data obtained from five mice

[b] Total anti-TNP PFC/spleen response was measured 5 days after immunization. The percentage M460 Id⁺ anti-TNP PFC was determined by the addition of anti-M460 Id antisera to the agarose and scoring the difference in the number of PFC without antisera. Results represent the mean ± SEM for the data obtained from five mice

with anti-E109 IdX antibodies before immunization (LIEBERMAN et al. 1979). This finding prompted HIERNAUX et al. (1981) to try to activate this silent clone in normal BALB/c mice. They showed that 1-day-old BALB/c mice treated with 0.01–10 μg BALB/c anti-A48 and immunized at 4, 5, or 7 weeks of age with BL produced anti-BL PFC which were 45%–73% A48 Id⁺ as compared with mice not treated with anti-A48, which expressed only 6% A48 Id⁺ PFC. These studies also showed that immunization with BL was required for the expression of A48.

The A48 system was examined in greater detail by RUBINSTEIN et al. (1983). These studies confirmed the earlier findings of HIERNAUX et al. (1981) and, in addition, demonstrated two important features of this system: (a) anti-Id can behave as an internal image of antigen, and (b) the effect of anti-Id priming can be transferred to normal mice by purified B cells.

In these experiments, mice were treated at birth with polyclonal BALB/c anti-A48 and subsequently were immunized with BL. The requirement for antigen following anti-Id treatment is shown in Table 1. Moreover, as presented in Table 2, a monoclonal anti-A48 could substitute for the requirement for BL challenge. Of six monoclonal anti-A48 antibodies, one, 17–38, was able to induce a BL PFC response in mice primed at birth with polyclonal anti-A48. Mice treated at birth with anti-A48 and given 17–38 at 1 month of age made substantial amounts of IgG anti-BL antibodies, as detected by IEF (Fig. 1), that were of the normal adult BALB/c spectrotype (STEIN et al. 1980) and that lacked the A48 Id. Mice given anti-A48 at birth and immunized with BL also produced IgG anti-BL, but these antibodies were of a different spectrotype and were A48 Id⁺. These data suggested that the 17–38 monoclonal antibody

Table 2. Ability of 17–38 monoclonal anti-A48 antibody to elicit an anti-BL PFC response in 12 mice treated at birth with 0.01 µg anti-A48 Id antibody (from RUBINSTEIN et al. 1983, by copyright permission of The Rockefeller University press)

Monoclonal anti-A48 antibody[a]	Isotype	Mean anti-BL PFC/spleen
26–125	IgG1	180
148–3	IgG1	140
100–105	IgG2b	180
23–113	IgM	120
14–114	IgM	40
17–38	IgM	2120

[a] Each mouse received a 10-µg injection of the appropriate monoclonal anti-A48 Id antibody, and the anti-BL PFC response was measured 5 days later. Results are expressed at the average of a mice per group

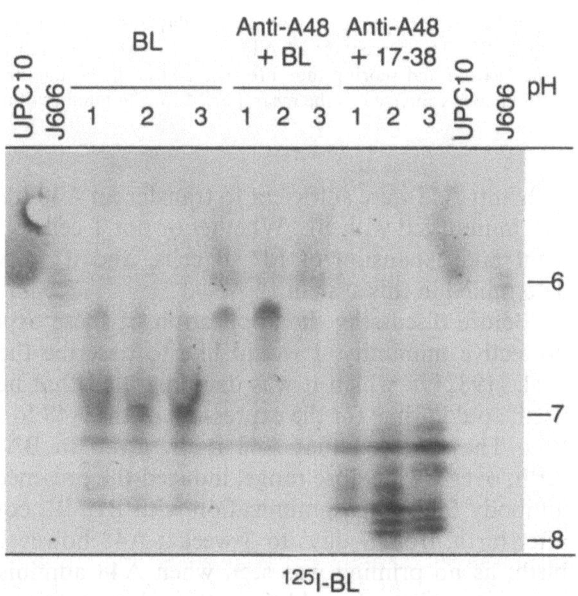

Fig. 1. A quantity of 10 µl serum were obtained 5 days after immunization from 1-month-old BALB/c mice and analyzed by IEF autoradiography using ^{125}I-labeled I BL antigen overlay. The groups are indicated *above* the autoradiogram and the individual sera are indicated by the *numbers* within each group. BL mice were not primed at birth and were given BL at 1 month; anti-A48 + BL mice were primed at birth and given BL at 1 month; anti-A48 + 17–38 mice were primed at birth with anti-A48 and given 17–38 at 1 month. For reference, 10 µl of a 1 mg/ml solution of U10 and J606 purified myeloma proteins were applied to the gel. The autoradiogram was obtained after exposing the film for 8 days (from RUBINSTEIN et al. 1983, by copyright permission of The Rockefeller University Press)

expressed two functional determinants, anti-A48 in its paratope (antigen-combining site), which suppressed the expression of A48, and an internal image of BL in its idiotope which stimulated the production of "normal" BALB/c anti-BL. This anti-Id, expressing an internal image of antigen, would be classified by JERNE et al. (1982) as an Ab2β (also see the chapter by D. Sacks in this volume).

A second important finding in these studies was the demonstration that the effect of neonatal priming was directly on the B cells. Table 3 shows that purified B cells, alone, taken at 1 month of age from mice primed at birth

Table 3. Transfer of A48 Id$^+$ response with B cells from mice treated at birth with 0.01 μg anti-A48 Id antibody (from RUBINSTEIN et al. 1983, by copyright permission of the The Rockefeller University press)

Lethally irradiated BALB/c mice infused with		Mice (*n*)	Anti-BL PFC/spleen[c]	
B cells[a]	T cells[b]		Total response (*n*)	A48 Id$^+$ (%)
Nil	Nil	2	180±170	7±7
Normal	Nil	2	843±375	0
Anti-A48	Nil	7	1857±964	46±14
Normal	Anti-A48	3	940±530	0
Anti-A48	Normal	7	1094±449	42±11
Anti-A48	Anti-A48	2	650±75	29±6

[a] 40×10^6 B cells isolated from either normal BALB/c mice or BALB/c mice treated at birth with 0.01 μg anti-A48 Id

[b] 20×10^6 T cells isolated from either normal BALB/c mice or BALB/c mice treated at birth with 0.01 μg anti-A48 Id

[c] Anti-BL PFC response was assayed 5 days after the infusion of B and/or T cells along with 20 μg BL. The percentage of A48 Id$^+$ PFC was determined by adding anti-A48 Id antisera to the agarose and scoring the difference beteween the number of plaques and the total response. Results are expressed as the mean ± SEM for the number of mice indicated

with anti-A48 were sufficient to transfer an A48 Id$^+$ response in adoptive recipients immunized with BL. Whether or not T cells play a role early in the development and expansion of Id$^+$ B cells, and if so, what role, has not yet been determined in this system.

Before discussing the application of these experiments to the induction of protective immunity, I would like to describe the experiments of RUBINSTEIN et al. (1982) in which it was demonstrated that neonatal administration of Id, itself, could prime for the expression of the A48 Id following antigen administration. They showed that A48 itself, given to BALB/c mice within 24 h after birth, over a wide dose range, induced the presence of A48 Id$^+$ PFC and serum antibody following immunization with BL. BL could be given at various times after birth, from 7 days to 3 weeks; A48 however, had to be given soon after birth, as no priming was seen when A48 administration was delayed to 7–28 days after birth. In addition, they demonstrated that the effect of neonatal Id priming could be adoptively transferred to recipients by Lyt-1$^+$ helper T cells but not by B cells. These studies strongly support others which reported the existence of Id-specific helper T cells (ThId, see BOTTOMLY 1984) and suggest an important role for maternal antibody in newborns in addition to passive protection.

3.3 Neonatal Priming for Protection Against *E. coli* K13 Infection

The above-described studies in the A48 system formed the basis for our experiments on the use of neonatal priming to induce protection against infection

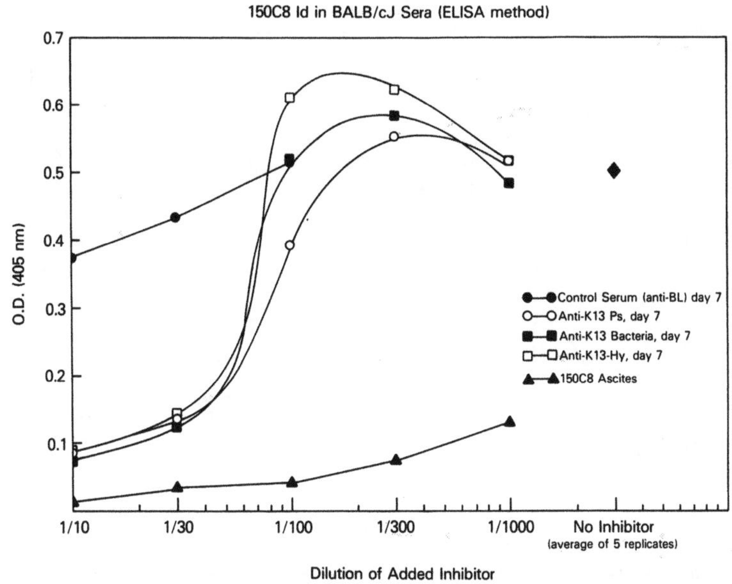

Fig. 2. Representative sera obtained from BALB/c mice 7 days after immunization with K13 polysaccharide (K13 PS), K13 whole cell vaccine (K13 bacteria), or K13 polysaccharide coupled to horseshoe hemocyanin (K13-Hy) were used to inhibit the binding of alkaline phosphatase-labeled 5868C anti-Id to 150C8 coated polyvinyl plates. Controls consisted of a serum obtained 7 days after immunization with bacterial levan (anti-BL) and 150C8 ascites. The *diamond* shows the optical density of the uninhibited reaction

with *E. coli* K13. The following three criteria are important for the selection of anti-Id to be used for the induction of protective immunity: (a) the Id should confer passive protection; (b) the Id should increase in response to immunization with antigen; (c) the anti-Id-Id interaction should be inhibited by antigen.

A series of monoclonal antibodies were prepared in BALB/c mice against the polysaccharide capsule of *E. coli* K13 (Soderstrom et al. 1983). One of these, 150C8, an IgM antibody, was shown to confer passive protection against infection with *E. coli* 06:K13:H1 (Soderstrom et al. 1982). The protein was used to prepare anti-Id antibodies in A/He mice (Stein and Soderstrom 1984). Anti-150C8 monoclonal antibodies were prepared by fusion of A/He immune spleen cells with the SP2/O cell line. An IgG1 anti-150C8, 5868C, was selected for further study. It was first used to evaluate the presence of the 150C8 Id following immunization with K13 immunogens. We found the 150C8 Id in pre-immune sera of BALB/c, CBA, and C57BL/6 mice and it increased in titer 7 days after immunization with K13 PS or whole cell vaccine in BALB/c and CBA sera, but not in C57BL/6 (Stein and Soderstrom 1984). This is shown for BALB/c sera in Fig. 2 as assayed by ELISA in which sera were used to inhibit the binding of alkaline phosphatase labeled 5868C to 150C8. In sera obtained 14 and 21 days after a primary immunization or 7 days after a secondary immunization, the level of 150C8 was below the pre-immune level and these sera failed to inhibit in this assay. Thus, in BALB/c and CBA mice,

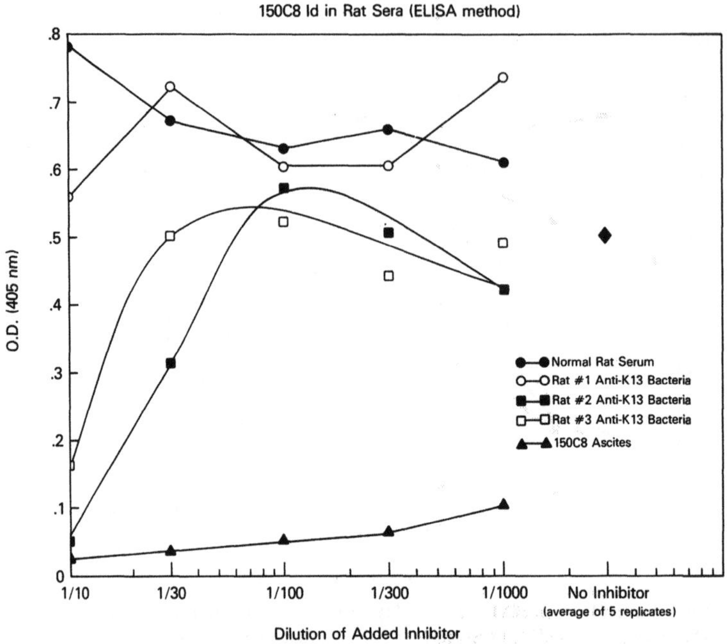

Fig. 3. Sera from three individual rats were obtained 1 week after the last of 6 monthly injections with a K13 whole cell vaccine. Controls consisted of normal rat serum (pool of the preimmune bleedings) and 150C8 ascites. The ELISA was performed as described in the legend of Fig. 2

150C8 is an early primary Id. In this regard it resembles the NP^b Id described by IMANISHI and MAKELA (1975). We were also able to detect 150C8 in hyperimmune sera from two of three Sprague-Dawley rats, and not in a pool of preimmune sera from these rats (Fig. 3, and STEIN and SODERSTROM 1984). Another group of rats had detectable levels of 150C8 in the preimmune sera and failed to show increases after immunization. Although the reason for the variation is not yet clear (these rats are outbred), 150C8, an early primarly Id in mice, can be detected in hyperimmune anti-K13 sera of some rats. Thus, 5868C would fit one of the criteria outlined by NISONOFF and LAMOYI (1981) for a candidate anti-Id containing a "related epitope," namely the interaction of anti-Id with anti-K13 in different strains and species.

It was stated that the anti-Id-Id interaction should be inhibited by antigen. If an anti-Id is going to be able to induce antibody of a given specificity (and also avoid inducing Id^+ antigen-nonbinding molecules) then it should interact with the antibody at its combining site and this interaction will, therefore, be inhibited by antigen. Table 4 shows that K13 PS inhibits the agglutination of 5868C coated erythrocytes by 150C8. Interestingly, although 150C8 does not precipitate directly with two closely related PS, K20 and K23, a weak cross-reactivity was detected by this very sensitive hemagglutination inhibition assay. Unrelated PS, K1, and Hib did not inhibit even at concentrations as high as 10 mg/ml.

Table 4. Inhibition of idiotype-anti-idiotype by antigen (from Stein and Soderstrom 1984, by copyright permission of The Rockefeller University Press)

	Polysaccharide[a]	HAI titer[b]
K13	3)-β-Ribofluranosyl-(1 → 7)-β-KDO-(2 → 4 ↑ Acetyl	44
K20	3)-β-Ribofuranosyl-(1 → 7)-β-KDO-(2 → 5 ↑ Acetyl	5
K23	3)-β-Ribofuranosyl-(1 → 7)-β-KDO-(2 →	6
K1	8)-α-NeuNAc-(2 → 8)-α-NeuNAc-(2 →	0
HIB	3)-β-Ribofuranosyl-(1 → 1)-Ribitol-5-(PO_4 →	0

[a] Polysaccharide solutions were used at a starting concentration of 10 mg/ml in saline, and serial two fold dilutions were used for inhibition
[b] log 2

Inasmuch as 5868C fulfilled the criteria that we established for an anti-Id that might prime for an anti-K13 response, we used it in an experimental system where BALB/c mice were injected within 24 h after birth with either anti-Id or Id in sterile phosphate-buffered saline. The mice were immunized at 4 weeks of age with either a K13 whole cell vaccine (experiment 1) or K13 PS (experiment 2) and 1 week later were challenged i.p. with 20 or 30 LD_{50} *E. coli* 06:K13:H1. The percent survival was evaluated 24 h after infection (Table 5). The results show that mice injected at birth with Id or anti-Id were protected from death caused by *E. coli* K13 infection, thus confirming the findings of Rubinstein et al. (1982, 1983) that both Id and anti-Id can prime for a response subsequent to antigen administration. In addition, anti-Id given to a mother within 24 h after delivery of her pups, and in a higher dose than given directly to the pups, was able to prime the pups via the milk (experiment 1). The effect of anti-Id priming was long lasting as shown in Table 6. When the immunization with K13 PS was delayed until the mice were 12 weeks old and then challenged 1 week later with 50 LD_{50} *E. coli* 06:K13:H1, 78% survived the challenge as compared with 25% of mice given only K13 PS. In this experiment the effect of Id priming was not seen but this needs to be further investigated.

It is important to point out that the primed mice were challenged with *E. coli* 1 week after immunization with K13. As we have described (Stein and Sonderstrom 1984) and as shown in Fig. 2, the 150C8 Id is at high levels on day 7 after a primary immunization, thus mice challenged 1 week after immunization should have had optimal levels of 150C8. Other systems may very well require different experimental conditions. For example, although A48 is a "silent" Id in BALB/c mice and is not normally expressed following immunization with BL, CBA/Ca mice do express A48 28 days after administration of BL (K.E. Stein, unpublished observations). In other words, A48 is a late primary Id, in contrast to 150C8.

Table 5. Effect of neonatal priming with idiotype and anti-idiotype when BALB/c mice are challenged at 5 weeks of age (from STEIN and SODERSTROM 1984, by copyright permission of The Rockefeller University Press)

Treatment at birth	Immunization at 4 weeks of age	Mice (*n*)	Survival (%)
Experiment 1			
–	–	6	0[a]
–	K13 vaccine	9	44
K13 Ps, 2.5 μg	K13 vaccine	6	33
150C8, 1 μg	K13 vaccine	15	87
5868C, 50 ng	K13 vaccine	15	93
5868C, 10 μg given to mothers after delivery[b]	K13 vaccine	6	83
Experiment 2			
Saline	K13 Ps, 2.5 μg	6	0[c]
K13 Ps, 2.5 μg	K13 Ps, 2.5 μg	6	33
150C8, 1 μg	K13 Ps, 2.5 μg	8	75
150C8, 50 ng	K13 Ps, 2.5 μg	10	40
5868C, 1 μg	K13 Ps, 2.5 μg	5	80
5868C, 50 ng	K13 Ps, 2.5 μg	9	78

–, no injection
[a] Mice were challenged with 20 LD_{50} live *E. coli* 06:K13
[b] Anti-idiotype was injected i.p. into the mother within 24 h after delivery of her pups
[c] Mice were challenged with 30 LD_{50} live *E. coli* 06:K13

Table 6. Effect of idiotype and anti-idiotype priming when BALB/c mice are challenged at 13 weeks of age (from STEIN and SODERSTROM 1984, by copyright permission of the Rockefeller University Press)

Treatment at birth	Immunization at 12 weeks of age	Mice (*n*)	survival[a] (%)
–	–	9	0
–	K13 Ps, 2.5 μg	20	25
150C8, 1 μg	–	5	0
150C8, 1 μg	K13 Ps, 2.5 μg	27	22
5868C, 1 μg	K13 Ps, 2.5 μg	19	16
5868C, 50 ng	K13 Ps, 2.5 μg	30	78

[a] Mice were challenged with 50 LD_{50} live *E. coli* 06:K13
[b] –, no injection

4 Adult Priming for Antipolysaccharide Responses with Anti-Id

4.1 Antigroup A Streptococcal Carbohydrate

Anti-Id anibodies have been shown to prime for anti-PS antibodies (or a hapten, PC, which is present on a PS) in adult mice. The first report was in a paper by Eichmann (1974) in which the experiments were designed to see if guinea pig anti-Id antibodies raised against the A5A Id, which is characteristic of the A/J response to the PS of group A streptococci, could suppress the A5A Id. He injected mice with IgG fractions of anti-A5A and 10 days later started a series of immunizations with group A streptococci. On day 25, the mice were bled and the sera examined for antigroup A carbohydrate antibodies and A5A Id levels. Eichmann found that the IgG2 fraction of Anti-A5A did suppress the A5A Id, however, the IgG1, noncomplement-fixing, fraction of anti-A5A, tended to enhance the level of A5A.

4.2 Anti-PC and Protective Immunity Against *Streptococcus Pneumoniae*

In mice, virtually all of the antibody made against the cell wall C polysaccharide (CPS) of *S. pneumoniae* reacts with its haptenic determinant, PC, (K.E. Stein, unpublished observations). Thus, although PC is not itself a polysaccharide, it is the immunodominant group of CPS and can be considered as one in terms of its immunogenic properties in mice. In 1975, Trenkner and Riblet reported the induction of anti-PC antibodies by the use of anti-S107 (a PC-binding mouse myeloma protein) made in rabbits and in A/He mice. They showed that anti-S107 was capable of inducing anti-PC PFC in vitro in a T dependent manner. The induction of anti-PC was not dependent on antigen, but antigen plus anti-Id has a synergistic effect and this could be shown both in vitro and in vivo.

Nearly 10 years later, McNamara et al. (1984) demonstrated that a monoclonal anti-T15 antibody, 4C11, when coupled to keyhole-limpet hemocyanin (KLH) can be used to immunize mice to produce not only anti-PC antibodies, but also protection against infection with *S. pneumoniae*. In their system, the induction of anti-PC antibodies did not require the administration of antigen.

5 Conclusions and Speculations

Two major findings emerge from the studies of anti-Id priming for antipolysaccharide responses in neonates: (a) administration of anti-Id antibodies to newborn mice results in the activation of Id^+ B cell clones that are subsequently responsive to antigen; in contrast, administration of polysaccharide antigens, themselves, to newborns does not stimulate antibody responses; (b) administration of monoclonal anti-Id antibodies, directed against a single epitope, can prime for protection against a lethal bacterial infection. Clearly, neonatal anti-Id priming can have profound effects on the immune system as has been demonstrated, but many aspects of this mode of inducing immunity remain to be clarified.

Can protein molecules structurally resemble carbohydrates? I have presented some data in this chapter that indicate that anti-Id can function as Ab2β (Jerne et al. 1982) in the induction of anti-PS antibodies although neither the structures of these anti-Id (17–38 and 5868C) nor the conformations in solution of the respective antigens, BL and K13, are known. Thus, although no proof exists, the fact that anti-Id antibodies can mimic carbohydrate determinants seems clear. Further evidence comes from the recent work of D. Sacks (see his chapter in this volume), in which a putative Ab2β is able to induce anticarbohydrate antibodies directed against a major glycoprotein antigen of *Trypanosoma cruzi*.

Another question to be answered is: Does anti-Id treatment of newborns prime Lyb5$^-$ cells directly? The studies of RUBINSTEIN et al. (1983) indicate that the effect of anti-Id priming can be transferred by B cells and these are presumed to be Lyb5$^-$ if they are primed at birth, before Lyb5$^+$ cells are detectable. A second indication that Lyb5$^-$ cells can be primed by anti-Id comes from studies in progress in my laboratory which show that anti-A48 can prime for anti-BL antibodies in CBA/N mice, which carry the *xid* gene defect, lack Lyb5$^+$ cells, and fail to respond to PS antigens. If anti-Id (Ab2) antibodies do prime Lyb5$^-$ cells directly then what is the mechanism of triggering by which Ab2 works but antigen does not?

What are the constraints on priming for anti-PS responses with Ab2 as compared with antiprotein responses? I raise this question because PS antigens are in the environment (both internal and external) all the time. Many of these PS antigens are cross-reactive. In contrast, for example, one is not constantly exposed to pathogenic viruses, the majority of protective antigens are protein, and exposure usually stimulates life-long antigen-specific T cell immunity. It is essential to understand the mechanism by which Ab2 primes for anti-PS responses and the potential problems before using such reagents as vaccines in children.

Finally, data reviewed in this chapter demonstrate that Id, itself, can prime newborns. This suggests that the role of maternal antibody may be not simply to provide passive protection to the newborn but also to prime Id-specific helper T cells (ThId) which can presumably interact with Id$^+$ B cells in concert with antigen to amplify the immune response. Although antibody responses to PS antigens are T independent, T cells clearly can modulate the response (MONGINI et al. 1981; MOND et al. 1983; BOTTOMLY 1984). Idiotype priming, because of its ability to stimulate ThId is, thus, an important protective mechanism in the neonate. Although the natural way to Id prime is via maternal antibody, we may soon see the use of synthetic idiotopes as immunogens. Recently, MCMILLAN et al. (1983) and SEIDEN et al. (1984) demonstrated that synthetic peptides corresponding to idiotopes on M104 and J558 (dextran-binding myeloma proteins) can can be used to immunize mice to produce anti-Id specific for the myeloma protein whose sequence was used to synthesize the peptide. Such synthetic idiotopes may be ideal immunogens to prime ThId.

Acknowledgments. I wish to acknowledge the valued collaboration of C. Bona and his graduate student, L. Rubinstein, and T. Soderstrom. The expert assistance of G. Connor in preparing this manuscript is greatly appreciated.

References

Ahmed A, Scher I, Scharrow SO, Smith AH, Paul WE, Sachs DH, Sell KW (1977) B-lymphocyte heterogeneity: development and characterization of an alloantiserum which distinguishes B-lymphocyte differentiation alloantigens. J Exp Med 145:101–110

Anderson P (1983) Antibody responses to *Haemophilus influenzae* type b and diphtheria toxin induced by conjugates of oligosaccharides of the type b capsule with the nontoxic protein CRM_{197}. Infect Immun 39:233–238

Anderson PW, Pichichero ME (1984) Antibody (Ab) to the capsular polysaccharide (PRP) of *Haemophilus influenzae* b in 2 to 6-mo-old infants vaccinated with PRP oligosaccharides conjugated to the diphtheria protein CRM_{197}. Pedriatr Res 18:252A

Baker PJ, Morse HC III, Cross SS, Stashak PW, Prescott B (1977) Maturation of regulatory factors influencing magnitude of antibody response to capsular polysaccharide type III *Streptococcus pneumoniae*. J Infect Dis 136:S20–S24

Band JD, Chamberland ME, Platt T, Weaver RE, Thornsberry C, Fraser DW (1983) Trends in meningococcal disease in the United States 1975–1980. J Infect Dis 148:754–758

Beuvery EC, Kaaden AVD, Kanhai V, Leussink AB (1983) Physicochemical and immunological characterization of meningococcal group A polysaccharide-tetanus toxoid conjugates prepared by two methods. Vaccine 1:31–36

Björksten B, Hattevig G, Kjellman B, Ahlstedt St, Richter W (1982) Dextran-reactive antibodies in healthy infants and toddlers – relation to type of feeding. Int Arch Allergy Appl Immunol 69:174–178

Bona C, Lieberman R, Chien CC, Mond J, House S, Green I, Paul WE (1978) Immune response to levan I. Kinetics and ontogeny of anti-levan and anti-inulin antibody response and of expression of cross-reactive idiotype. J Immunol 120:1436–1442

Bona C, Mond JJ, Stein KE, House S, Lieberman R, Paul WE (1979) Immune response to levan III. The capacity to produce anti-inulin antibodies and cross-reactive idiotypes appears late in ontogeny. J Immunol 123:1484–1490

Bottomly K (1984): all idiotypes are equal, but some are more equal than others. Immunol Rev 79:45–61

Briles DE, Davie JM (1975) Clonal Dominance I. Restricted nature of the IgM antibody response to group A streptococcal carbohydrate in mice. J Exp Med 141:1291–1307

Briles DE, Forman C, Hudak S, Claflin JL (1982) Anti-phosphorylcholine antibodies of the T15 idiotype are optimally protective against *Streptococcus pneumoniae*. J Exp Med 156:1177–1185

Briles DE, Forman C, Hudak S, Claflin JL (1984) The effects of idiotype on the ability of IgG_1 anti-phosphocholine antibodies to protect mice from fatal infection with *Streptococcus pneumoniae*. Eur J Immunol 14:1027–1030

Cancro MP, Sigal NH, Klinman NR (1978) Differential expression of an equivalent clonotype among BALB/c and C57BL/6 mice. J Exp Med 147:1–12

Chu C, Schneerson R, Robbins JB, Rastogi SC (1983) Further studies on the immunogenicity of *Haemophilus influenzae* type b and pneumococcal type 6A polysaccharide-protein conjugates. Infect Immun 40:245–256

Cramer M, Braun DG (1975) Genetics of restricted antibodies to streptococcal group polysaccharides in mice II. The Ir-A-CHO gene determines antibody levels, and regulatory genes influence the restriction of the response. Eur J Immunol 5:823–830

Der Balian GP, Slack J, Clevinger BL, Bazin H, Davie JM (1980) Subclass restriction of murine antibodies III. Antigens that stimulate IgG_3 in mice stimulate IgG_{2c} in rats. J Exp Med 152:209–218

Douglas RM, Miles HB (1984) Vaccination against *Streptococcus pneumoniae* in childhood: lack of demonstrable benefit in young Australian children. J Infect Dis 149:861–869

Egan ML, Pritchard DG, Dillon HC Jr, Gray BM (1983) Protection of mice from experimental infection with type III group B streptococcus using monoclonal antibodies. J Exp Med 158:1006–1011

Eichmann K (1974) Idiotype suppression I. Influence of the dose and of the effector functions of anti-idiotypic antibody on the production of an idiotype. Eur J Immunol 4:296–302

Fernandez C, Moller G (1979) Antigen-induced strain-specific autoantiidiotypic antibodies modulate the immune response to dextran B512. Proc Natl Acad Sci USA 76:5944–5947

Fernandez C, Moller G (1980) A primary immune response to dextran B512 is followed by a period of antigen-specific immunosuppression caused by autoanti-idiotypic antibodies. Scand J Immunol 11:53–62

Geha RS (1983) Presence of circulating anti-idiotype-bearing cells after booster immunization with tetanus toxoid (TT) and inhibition of anti-TT antibody synthesis by auto-anti-idiotypic antibody. J Immunol 130:1634–1639

Gold R, Lepow ML, Goldschneider I, Gotschlich EC (1977) Immune response of human infants to polysaccharide vaccines of groups A and C *Neisseria meningitidis*. J Infect Dis 136:S31–S35

Golding B, Muchmore AV, Blaese RM (1984) Newborn and Wiskott-Aldrich patient B cells can be activated by TNP-*Brucella abortus:* evidence that TNP-*Brucella abortus* behaves as a T-independent type 1 antigen in humans. J immunol 133:2966–2971

Granoff DM, Boies EG, Munson RS Jr (1984) Immunogenicity of *Haemophilus influenzae* type b polysaccharide-diphtheria toxoid conjugate vaccine in adults. J Pediatr 105:22–27

Hansburg D, Briles DE, Davie JM (1976) Analysis of the diversity of murine antibodies to dextran B1355 I. Generation of a large, pauci-clonal response by a bacterial vaccine. J Immunol 117:569–575

Hansburg D, Clevinger B, Perlmutter RM, Griffith R, Briles DE, Davie JM (1979) Analysis of the diversity of murine antibodies to $\alpha(1 \rightarrow 3)$ dextran. In: Cells of immunoglobulin synthesis, B. Pernis and H.J. Vogel, eds. Academic, New York, pp 295–308

Hiernaux J, Bona C, Baker PJ (1981) Neonatal treatment with low codes of anti-idiotypic antibody leads to the expression of a silent clone. J Exp Med 153:1004–1008

Huber B, Gershon RK, Cantor H (1977) Identification of a B-cell surface structure involved in antigen-dependent triggering: absence of this structure on B cells from CBA/N mutant mice. J Exp Med 145:10–20

Imanishi T, Makela O (1975) Inheritance of antibody specificity. II. Anti-(4-hydroxy-5-bromo-3-nitrophenyl)acetyl in the mouse. J Exp Med 141:840–854

Jennings HJ, Lugowski C (1981) Immunochemistry of groups A, B, and C meningococcal polysaccharide-tetanus toxoid conjugates. J Immunol 127:1011–1018

Jennings HJ, Lugowski C, Ashton FE (1984) Conjugation of meningococcal lipopolysaccharide R-type oligosaccharides to tetanus toxoid as route to a potential vaccine against group B *Neisseria meningitidis*. Infect Immun 43:407–412

Jerne NK, Roland J, Cazenave PA (1982) Recurrent idiotopes and internal images. EMBO 1:243–247

Jorbeck HJA, Svenson SB, Lindberg AA (1981) Artificial *Salmonella* vaccines: *Salmonella typhimurium* O-antigen-specific oligosaccharide-protein conjugates elicit opsonizing antibodies that enhance phagocytosis. Infect Immun 32:497–502

Kaijser B, Alstedt S (1977) Protective capacity of antibodies against *Escherichia coli* O and K antigens. Infect Immun 17:286–289

Kayhty H, Karanko V, Peltola H, Makela PH (1984) Serum antibodies after vaccination with *Haemophilus influenzae* type b capsular polysaccharide and responses to reimmunization: no evidence of immunologic tolerance or memory. Pediatrics 74:857–865

Kunkel HG, Mannik M, Williams RC (1963) Individual antigenic specificity of isolated antibodies. Science 140:1218–1219

Lepow ML, Samuelson JS, Gordon LK (1985) Safety and immunogenicity of *Haemophilus influenzae* type b polysaccharide diphtheria toxoid conjugate vaccine (PRP-D) in infants 9–15 months. J Pediatr 106:185–189

Lieberman R, Potter M, Humphrey W Jr, Mushinski EB, Vrana M (1975) Multiple individual and cross-specific idiotypes on 13 levan-binding myeloma proteins of BALB/c mice. J Exp Med 142:106–119

Lieberman R, Bona C, Chien CC, Stein KE, Paul WE (1979) Genetic and cellular regulation of the expression of specific antibody idiotypes in the anti-polyfructosan immune response. Ann Immunol (Paris) 130C:247–262

Makela PH, Peltola H, Kayhty H, Jousimies H, Pettay O, Ruoslashti E, Sivonen A, Renkonen OV (1977) Polysaccharide vaccine of group A *Neisseria meningitidis* and *Haemophilus influenzae* type b: a field trial in Finland. J Infect Dis 136:S43–S50

Makela PH, Sibakov M, Herva E, Henrichsen J, Luotonen J, Timonen J, Leinonen M, Koskela M, Pukander J, Pontynen S, Gronroos P, Karma P (1980) Pneumococcal vaccine and otitis media. Lancet II:547–551

McMillan S, Seiden MV, Houghten RA, Clevinger B, Davie JM, Lerner RA (1983) Synthetic idiotypes: the third hypervariable region of murine anti-dextran antibodies. Cell 35:859–863

McNamara MK, Ward RE, Kohler H (1984) Monoclonal idiotype vaccine against *Streptococcus pneumonia* infection. Science 226:1325–1326

Mond JJ, Farrar J, Paul WE, Fuller-Farrar J, Schaefer M, Howard M (1983) T cell dependence and factor reconstitution of *in vitro* antibody responses to TNP-*B. abortus* and TNP-Ficoll: restoration of depleted responses with chromatographed fractions of a T cell-derived factor. J Immunol 131:633–637

Mongini PKA, Stein KE, Paul WE (1981) T cell regulation of IgG subclass antibody production in response to T-independent antigens. J Exp Med 153:1–12

Mosier DE, Zaldivar NM, Goldings E, Mond J, Scher I, Paul WE (1977a) Formation of antibody in the newborn mouse: study of T-cell-independent antibody response. J Infect Dis 136:S14–S19

Mosier DE, Zitron IM, Mond JJ, Ahmed A, Scher I, Paul WE (1977b) Surface immunoglobulin D as a functional receptor for a subclass of B lymphocytes. Immunol Rev 27:89–104

Nishinarita S, Claflin JL, Lieberman R (1985) IgA isotype restricted idiotypes associated with T15 Id^+ PC antibodies. J Immunol 134:2549

Nisonoff A, Lamoyi E (1981) Hypothesis. Implications of the presence of an internal image of the antigen in anti-idiotypic antibodies: possible application to vaccine production. Clin Immunol Immunopathol 21:397–406

Parke JC Jr, Schneerson R, Robbins JB, Schleselman JJ (1977) Interim report of a controlled field trial of immunization with capsular polysaccharides of *Haemophilus influenzae* type b and group C *Neisseria meningitidis* in Mecklenburg County, North Carolina (March 1974–March 1976). J Infect Dis 136:S51–S56

Peltola H, Kayhty H, Sivonen A, Makela PH (1977) *Haemophilus influenzae* in children: a double-blind field study of 100,000 vaccinees 3 months to 5 years of age in Finland. Pediatrics 60:730–737

Perlmutter RM, Briles DE, Davie JM (1977) Complete sharing of light chain spectrotypes by murine IgM and IgG anti-streptococcal antibodies. J Immunol 118:2161–2166

Perlmutter RM, Hansburg D, Briles DE, Nicolotti RA, Davie JM (1978) Subclass restriction of murine anti-carbohydrate antibodies. J Immunol 121:566–572

Pincus DJ, Morrison D, Andrews C, Lawrence E, Sell SH, Wright PF (1982) Age-related response to two *Haemophilus influenzae* type b vaccines. J Pediatr 100:197–201

Rodkey LS, Binion SB, Brown JC, Adler FL (1983) Natural regulation of antibody clones by auto-anti-idiotypic antibodies in rabbits. In: Bona CA, Kohler H (eds) Immune networks. Ann NY Acad Sci 418:16–25

Rubinstein LJ, Yeh M, Bona CA (1982) Idiotype-anti-idiotype network. II. Activation of silent clones by treatment at birth with idiotypes is associated with the expansion of idiotype specific helper T cells. J Exp Med 156:506–521

Rubinstein LJ, Goldberg B, Hiernaux J, Stein KE, Bona CA (1983) Idiotype-antiidiotype regulation V. The requirement for immunization with antigen or monoclonal antiidiotypic antibodies for the activation of $\beta2 \rightarrow 6$ and $\beta2 \rightarrow 1$ polyfructosan-reactive clones in BALB/c mice treated at birth with minute amounts of anti-A48 idiotype antibodies. J Exp Med 158:1129–1144

Sacks DL, Kelsoe GH, Sachs DH (1983) Induction of immune responses with anti-idiotypic antibodies: implications for the induction of protective immunity. Springer Semin Immunopathol 6:79–97

Saxon A, Barnett E (1984) Human auto-antiidiotypes regulating T cell-mediated reactivity to tetanus toxoid. J Clin Invest 73:342–348

Schneerson R, Rodrigues LP, Parke JC Jr, Robbins JB (1971) Immunity to disease caused by *Hemophilus influenzae* type b. II. Specificity and some biologic characteristics of "natural," infection-acquired, and immunization-induced antibodies to the capsular polysaccharide of *Hemophilus influenzae* type b. J Immunol 107:1081–1089

Schneerson R, Barrera O, Sutton A, Robbins JB (1980) Preparation, characterization, and immunogenicity of *Haemophilus influenzae* type b polysaccharide-protein conjugates. J Exp Med 152:361–376

Schneerson R, Robbins JB, Chu C, Sutton A, Vann W, Vickers JC, London WT, Curfman B, Hardegree MC, Shiloach J, Rastogi SC (1984) Serum antibody responses of juvenile and infant rhesus monkeys injected with *Haemophilus influenzae* type b and pneumococcus type 6A capsular polysaccharide-protein conjugates. Infect Immun 45:582–591

Schrater AF, Goidl EA, Thorbecke GJ, Siskind GW (1979) Production of auto-anti-idiotypic antibody during the normal immune response to TNP-Ficoll I. Occurrence in AKR/J and BALB/c mice of hapten-augmentable, anti-TNP plaque-forming cells and their accelerated appearance in recipients of immune spleen cells. J Exp Med 150:138–153

Seiden MV, Clevinger B, McMillan S, Srouji A, Lerner R, Davie JM (1984) Chemical synthesis of idiotopes. Evidence that antisera to the same JH_1 peptide detect multiple binding site-associated idiotopes. J Exp Med 159:1338–1350

Smith DH, Peter G, Ingram DL, Harding AL, Anderson P (1973) Responses of children immunized with the capsular polysaccharide of *Hemophilus influenzae* type b. Pediatrics 52:637–644

Soderstrom T, Brinton CC Jr, Fusco P, Karpas A, Ahlstedt S, Stein K, Sutton A, Hosea S, Schneerson R, Hanson AL (1982) Analysis of pilus-mediated pathogenic mechanisms with monoclonal antibodies. In: Schlessinger D (ed) Microbiology – 1982. American Society for Microbiology, Washington DC, pp 305–307

Soderstrom T, Stein K, Brinton CC Jr, Hosea S, Burch C, Hansson HA, Karpas A, Schneerson R, Sutton A, Vann WI, Hanson LA (1983) Serological and functional properties of monoclonal antibodies to Escherichia coli Type 1 pilus and capsular antigens. Prog Allergy 33:259–274

Stein KE, Soderstrom T (1984) Neonatal administration of idiotype or antiidiotype primes for protection against *Escherichia coli* K13 infection in mice. J Exp Med 160:1001–1011

Stein KE, Bona C, Lieberman R, Chien CC, Paul WE (1980) Regulation of the anti-inulin antibody response by a nonallotype-linked gene. J Exp Med 151:1088–1102

Stein KE, Zopf DA, Johnson BM, Miller CB, Paul WE (1982) The immune response to an isomaltohexosyl-protein conjugate, a thymus-dependent analogue of $\alpha(1\rightarrow6)$ dexran. J Immunol 128:1350–1354

Stein KE, Zopf DA, Miller CB, Johnson BM, Mongini PKA, Ahmed A, Paul WE (1983) Immune response to a thymus-dependent form of B512 dextran requires the presence of Lyb-5^+ lymphocytes. J Exp Med 157:657–666

Svenson SB, Lindberg AA (1981) Artificial *Salmonella* vaccines: *Salmonella typhimurium* O-antigen-specific oligosaccharide-protein conjugates elicit protective antibodies in rabbits and mice. Infect Immun 32:490–496

Trenkner E, Riblet R (1975) Induction of antiphosphorylcholine antibody formation by anti-idiotypic antibodies. J Exp Med 142:1121–1132

Vodopija I, Baklaic Z, Hauser P, Roelants P, Andre FE, Safary A (1983) Reactivity and immunogenicity of bivalent (AC) and tetravalent ($ACW_{135}Y$) meningococcal vaccines containing O-acetyl-negative or O-acetyl-positive group C polysaccharide. Infect Immun 42:599–604

Ward J, Berkowitz C, Pescetti J, Burkart K, Samuelson J, Gordon L (1984) Enhanced immunogenicity in young infants of a new *Haemophilus influenzae* type b (HIB) capsular polysaccharide (PRP)-diphtheria toxoid (D) conjugate vaccine. Pediatr Res 18:287A

Winkler M, Urbain J (1984) Idiotypic manipulation of the rabbit immune response against *Micrococcus luteus*. In: Kohler H, Urbain J, Cazenave PA (eds) Idiotypy in biology and medicine. Academic, New York, pp 219–241

Yount WJ, Dorner MM, Kunkel HG, Kabat EA (1968) Studies on human antibodies VI. Selective variations in subgroup composition and genetic markers. J Exp Med 127:633–646

Zahradnik JM, Gordon LK (1984) Augmented antibody (Ab) responses in infants administered a new *Haemophilus influenzae* type b cappsular polysaccharide (PRP) diphtheria toxoid conjugate vaccine (PRP-D). Pediatr Res 18:289A

Human Antibody Induction to the Idiotypic and Anti-Idiotypic Determinants of a Monoclonal Antibody Against a Gastrointestinal Carcinoma Antigen

E. DeFreitas, H. Suzuki, D. Herlyn, M. Lubeck, H. Sears, M. Herlyn, and H. Koprowski

1 Introduction

Murine monoclonal antibodies (Mab) generated against human tumor antigens (Koprowski and Steplewski 1982) have renewed scientific and clinical interest in passive immunotherapy of human cancer (Kennett et al. 1980). Treatment of B cell leukemias with Mab, for instance, results in immediate reduction of leukemic cells in circulation (Levy and Miller 1983; Ritz et al. 1981). A Mab (17-1A) against an antigen of gastrointestinal cancers (Herlyn et al. 1979) was found to suppress the growth of human tumors in nude mice (Herlyn et al. 1980) and produced significant improvement in the clinical status of a number of patients with metastatic disease (Koprowski 1984a; Sears et al. 1984). As expected, most patients injected with this Mab produced antibody against the murine immunoglobulin (Ig) but, more germaine to this forum, a significant proportion of this antibody was directed against the antigen binding or idiotypic (Id) portion of the molecule (anti-Id) (Koprowski et al. 1984).

The importance of antigen receptor-specific antibody in the regulation of the immune response was recognized by Jerne (1974) and set forth as the "network theory of the immune system." This theory proposes that each member of the immune system interacts with others via the antigenic determinants en-

The Wistar Institute of Anatomy and Biology, 36th Street at Spruce, Philadelphia, PA 19104, USA

Current Topics in Microbiology and Immunology, Vol. 119

coded by variable regions (V) genes which are expressed both on immunoglobulin and antigen-reactive lymphocytes. Direct demonstration of this regulation has been shown by injection of specific Ig into syngeneic, allogeneic, or xenogeneic hosts which produce (among others) anti-Id antibodies (CAZANAVE 1977; FERNANDEZ and MOLLER 1979; GOIDL et al. 1979). These anti-Id antibodies are comprised of several populations of antibodies, designated as (BONA and KOHLER 1984): Ab2α, those which recognize antigen-noninhibitable idiotopes; Ab2γ, those which recognize antigen-inhibitable idioitopes; and Ab2β, those which interact with the antigen-receptor on specific Ig, B cells, and regulatory T-cell clones. Ab2β are therefore "internal images of the antigen" (LINDEMANN 1979) and are able to induce synthesis of Ab3 which share idiotopes and the antigen-binding properties of Ab1 and therefore modulate the response to the priming antigen (URBAIN et al. 1977).

A prediction of this network theory is that immunization with Ab2β would prime an animal to subsequent challenge with antigen. This has been elegantly demonstrated in several experimental systems. SEGE and PETERSON (1978) showed that Ab2 against insulin-binding Ab1 idiotope mimicked the action of insulin. KENNEDY et al. (1984) induced antibody to hepatitis B virus in mice by injection of xenogeneic antibodies against an Id of HB-s binding immunoglobulin. REAGAN et al. (1983) elicited rabies virus-neutralizing antibody in mice primed with anti-Id xenogeneic antibodies prepared against Mab specific for rabies virus glycoprotein.

It is conceivable, therefore, that passive immunotherapy of cancer patients with tumor antigen-specific Mab may induce "active" immunization against the antigen-bearing tumor cells via the induction of Ab2β. This may especially be relevant when no immediate antitumor effect of Mab is observed but occurs several months after the administration of Mab (MILLER et al. 1982; KOPROWSKI 1984a).

2 Induction of Human Antimurine and Anti-Id Antibodies After Immunization with 17-1A Mab (Ab1)

Selection of the Mab 17-1A for use in immunotherapy trials was based on its binding specificity on tumors of the gastrointestinal (GI) tract (HERLYN et al. 1979; KOPROWSKI et al. 1979), density of the antigen on (GI) tumor cells (KOPROWSKI 1984b; POWE et al. 1985), its ability to locate tumors by radioimaging (MOLDOFSKY et al. 1984; MACH et al. 1983), and its tumoricidal activity in conjunction with murine (HERLYN et al. 1980; HERLYN and KOPROWSKI 1982) and human (STEPLEWSKI et al. 1983) effector cells.

Immunotherapy trials with 17-1A have been completed on a group of 51 patients with gastrointestinal carcinoma (GIC) who had previously failed a trial of chemotherapy and had demonstrable metastases. Twenty-six patients developed an antimurine antibody response, and the majority of the patients who showed such response were found to develop anti-17-1A anti-Id. This was demonstrated in a radioimmunoassay (RIA) in which 17-1A Mab was absorbed on a bead, followed by patient's serum; the mixture was incubated overnight at 4° C, the beads thoroughly washed and exposed to ^{125}I-labeled goat anti-Id obtained after immunization of the animal with repeated injections of 17-1A Mab in complete Freund's adjuvant. The goat anti-Id was extensively absorbed to murine IgG2a Mabs other than 17-1A. A negative control of the anti-Id assay was provided by the use of anti-Id against Mab specific for colorectal cancer (CRC) antigen other than 17-1A.

Ten patients injected with Mab showed significant improvement in their disease status and were available for further study. Of these, six received a single injection of 17-1A and four received multiple injections. Antimurine Ig response was detected in nine patients (one was unavailable for study), and was, in most cases, demosntrated within 2 weeks of 17-1A administration. Disappearance of murine Ig from the circulation occurred between 5 and 50 days, depending on whether the Mab injection was a primary or secondary administration.

2.1 Human Ig Against 17-1A Mab: In Vivo Response

The majority of patients who developed an antimurine Ig response also developed anti-Id. The results of the assay on several consecutive serum samples from five patients are shown in Table 1. The results indicate that sera from patients 07, 08, 09, 14, and 23 significantly competed with heterologous anti-Id for binding to 17-1A Mab. Binding inhibition observed with serum by patient 14 was low but the values differed significantly from those obtained with pre-Mab exposure serum.

2.2 What Factors Determine Who Will Produce Anti-Id After 17-1A Injection?

The data from initial clinical trials with 17-1A Mab was analyzed to determine what factors might be involved in the anti-Id response of CRC patients. Although preformed natural antibodies to mouse Ig were not detected in pretherapy bleeds of any patient, it was conceivable that the 17-1A Mab-defined antigen on the patients' tumors primed Id-positive B-cell clones which, in turn, induced anti-Id B cells. Therefore, we chose to evaluate the presence of anti-Id positive B cells in the circulation of CRC patients immediately before 17-1A Mab therapy to determine whether patients with these cells were more likely to produce anti-17-1A antibody or to develop this antibody earlier.

Moreover, since the dose of antigen in experimental models appears to regulate development of active immunity or tolerance, this phase II clinical trial was designed wherein patients initially received a high dose (700 mg) or a low dose (100 mg) of 17-1A Mab in vivo followed by biweekly injections of 100 mg until a human antimurine Ig response developed.

During the course of Mab therapy, serial serum samples from each patient were tested for the presence of antimouse Ig by an RIA using 17-1A Mab as antigen and ^{125}I-labeled rabbit antihuman F(ab$'$)$_2$ (heavy and light chain-

Table 1. Inhibition of binding of Mab 17-1A to goat anti-Id 17-1A by sera of CRC patients

Patient	Days after Mab injection[a]	Inhibition of goat αId 17-1A binding to 17-1A[b]
07	0	0
	22	+
	58	+
	120	+
	178	+
	215	+
	475	+
	47	+
08	0	0
	13	+
	15	+
	17	+
	249	+
	598	+
	603	+
	605	+
	612	+
	668	+
09	0	0
	28	+
	238	+
	250	+
14	0	0
	135	+
23	0	0
	24	+
	66	+
	136	+
	337	+

[a] All patients received one administration of 17-1A Mab on day 0, except patient 08 who received a second injection on day 259

[b] Mab 17-1A (2.0 μg/ml) was coupled to polystyrene beads. Beads were incubated with several serum dilutions at 4° C overnight. Percentage of inhibition of binding of ^{125}I-labeled goat αId 17-1A to 17-1A was calculated by the following formula: % binding inhibition = 100 − [(100 × cpm bound in presence of serum)/cpm bound without serum]. Serum was considered positive when inhibition values were significantly ($P < 0.05$) higher than the values obtained with preinjection sera

specific) as the indicator (SEARS et al. 1984). Within the sensitivity of the assay, 16 of 19 patients produced an antimouse antibody response.

To correlate antibody responsiveness with the pre-existence of anti-17-1A Id-positive B cells, patients were grouped, as shown in Table 2, into those with such cells and those without. (Two patients were excluded because of the low yield of purified PB-B cells.)

We found no statistically significant difference between these two groups with respect to the proportion of people who made anti-17-1A antibody nor in the average time of onset of this response. This suggests that other regulatory

Table 2. Correlation between the pre-existence of anti-17-A Id-positive B cells and development of serum antimurine antibody

Patient[a]	PB-B cells binding Mab F(ab')2		Serum antibody response (day of first detection)	Average time of antibody onset (days)
	17-1A	A5C3		
46	3.8	0.0	21	22.3
48	2.6	0.0	7	
60	1.1	1.2	13	
61	0.8	0.0	14	
58	0.7	0.0	14	
47	0.6	0.0	14	
44	0.6	0.1	4	
49	0.3	0.4	0	
50	0.3	0.2	14	
54	0.2	0.2	99	
52	0.1	0.2	23	
57	0.1	0.0	0	
51	0.0	1.8	7	30.0
53	0.0	0.0	73	
55	0.0	0.0	21	
56	0.0	0.4	35	
59	0.0	0.0	14	

[a] PBMC were isolated from patients' blood immediately before 17-1A Mab immunotherapy. B cells were purified using density centrifugation and negative selection with $SE_{A\epsilon T}$ (SAXON et al. 1982). Cells were incubated with 17-1A or A5C3 (F(ab')2 and counterstained with F(ab')2 goat antimouse Ig-FITC. All populations were analyzed by flow cytometry on an Orthocytofluorograf. Monocytes were excluded based on sizing in the cytofluorograf

factors besides the presence of primed anti-17-1A Id-positive B cells in the circulation contribute to the development of an antimurine antibody response.

Of interest were two patients (51 and 60) who had a high proportion of A5C3 F(ab')2 binding B cells. Since this Mab detects a hepatitis B surface antigen (WANDS et al. 1982), we retrospectively evaluated patient 51 for hepatitis B S antigen serum antibody and found his titer to be 1:160 (patient 60 is currently being evaluated). Taken together, these data indicate that in vivo priming with antigen (17-1A antigen in CRC and hepatitis B surface antigen in the case of patient 51) induces Ab2-positive B cells most probably through stimulation with Ab1 and Ab1-positive B cells, but that the presence of these cells in circulation alone is not predictive of an antibody response.

A second parameter examined for its effect on the development of human anti-17-1A is the dose of Mab administered. This study was done with 19 patients from the immunotherapy group. Ten patients received 700 mg 17-1A initially and nine patients received 100 mg 17-1A initially. Both groups were treated with 100 mg 17-1A biweekly until human antimurine Ig response developed. Table 3 shows the patients in the study grouped according to their initial Mab dose. All of the nine patients receiving 100 mg initially responded with an early average onset of anti-17-1A antibody of 13.7 days, while of the ten patients who received 700 mg initially, three did not develop antimouse anti-

Table 3. Correlation between development of human α17-1A Mab and dose of Mab

Initial 17-1A Mab dose (mg)	Patient	Anti-17-1A Mab Serum Antibody		
		Response (day first detected)	Percent responders	Average onset (day)
700	45	0	70	37.6
	47	+(14)		
	49	0		
	51	+ (7)		
	53	+(73)		
	54	+(99)		
	55	+(21)		
	56	+(35)		
	57	0		
	58	+(14)		
100	44	+ (4)	100	13.7
	46	+(21)		
	48	+/−(7)		
	50	+(14)		
	52	+(23)		
	59	+(14)		
	60	+(13)		
	61	+(14)		
	62	+(13)		

body, and those that did showed an average onset of response of as late as 37.6 days. These results suggest that higher doses of murine Mab are not as immunogenic as lower doses, perhaps due to the induction of "high dose" tolerance as was demonstrated in the murine response to protein antigens.

2.3 Modulation of Regulatory T Cells After 17-1A Mab Immunotherapy

The proportions of phenotypic T helper (T_H) and T suppressor (T_S) cells in the peripheral blood were evaluated for all patients in the latest clinical trial. Since the ratio of T_H to T_S cells has been used empirically as a yardstick of general immune reactivity, these values were calculated for patients who were injected with 100 mg and developed an antimurine antibody response and for those who were injected with 700 mg (see above) who did not develop the response. Data for three representative patients in each group are shown in Fig. 1. Panel A (responders) shows that the T_H:T_S ratio fluctuates markedly in the course of therapy. The appearance of antibody (indicated by arrows) in the serum is preceded by dramatic rise of this ratio. Analysis of the absolute numbers of each type of cell indicates that this rise reflects *both* an increase in the number of T_H and a decrease in T_S. In contrast, the antibody nonresponders showed a decrease (or no change) in the T_H:T_S ratio. This pattern was observed regardless of the pretherapy ratio of these patients.

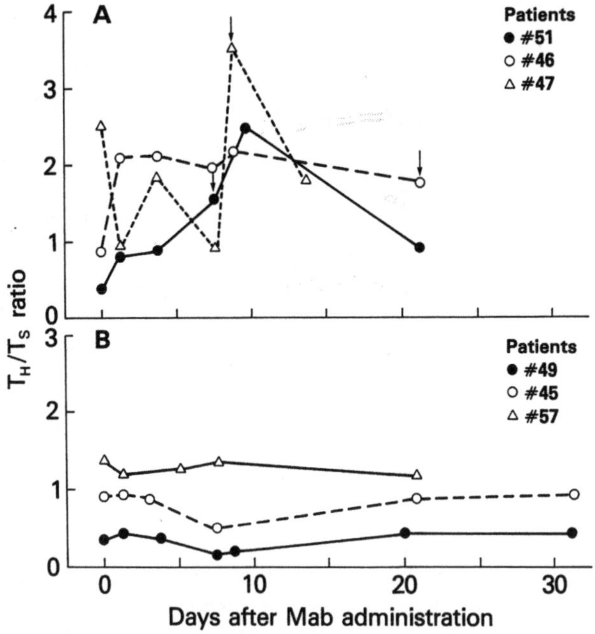

Fig. 1. T_H:T_S ratio in patients receiving 17-1A Mab treatment. PBMC were collected from GIC patients at the times indicated and analyzed for the percentage of OKT-4 positive (T_H) and OKT-8 positive (T_S) cells by flow cytometry. Panel **A** shows the data from patients who developed antimurine Ig antibody while panel **B** illustrates the ratios in antibody nonresponders. *Arrows* indicate day on which anti-17-1A antibody was detected in patients' serum

The major shifts in these populations of regulatory cells do not reflect the modulation of only antigen-specific T cells since this population most likely represents a minute fraction of the total T cell shifts measured. However, antigenic (17-1A Mab, in this case) stimulation is known to activate clones of cells of unrelated specificities. Subsets of these clones engage in mutual regulatory interactions defined as an "antigen-driven horizontal network" (GROSSMAN 1982) and the magnitude of the immune response generated was found to be correlated with the gradient of these cellular vascillations (GROSSMAN et al. 1980).

2.4 Induction of Human Anti-17-1A Antibody In Vitro from Peripheral Blood Lymphocytes of 17-1A Mab-Treated Patients

To determine the antigenic requirements and specificity of the human antimurine Ig response in CRC patients given 17-1A Mab, we examined the ability of their peripheral blood mononuclear cells (PBMC) to produce antibody after stimulation in a modified Mishell-Dutton culture system (DEFREITAS et al. 1982) using 17-1A F(ab')2 or F(ab')2 of an idiotypically unrelated γ2a Mab, A5C3. Representative results are shown in Fig. 2 with patients 23 and 8. PBMC from patient 23 was obtained 5 months after Mab treatment and from patient 8, 21 months after the first Mab treatment. After stimulation for 6 days with optimal concentrations (20 ng/ml) of 17-1A F(ab')2, both patients produced IgG against both 17-1A and A5C3. Stimulation with A5C3 F(ab')2 induced significantly less antibody to both Mab.

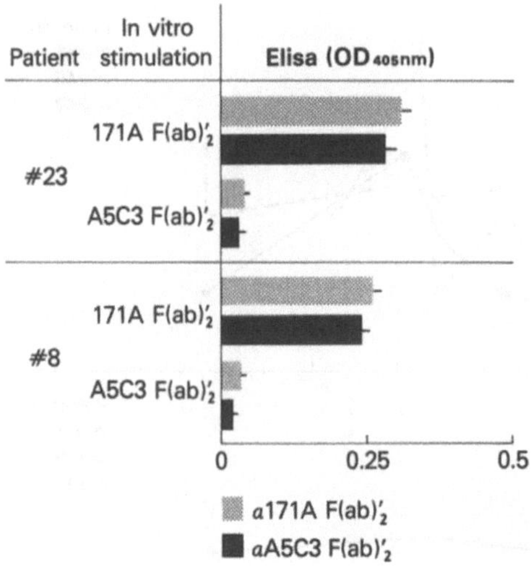

Fig. 2. Specificity of antibody produced in vitro by patients 23 and 8 after treatment with 17-1A Mab. PBMC from both patients were stimulated with either 17-1A F(ab')2 or A5C3 F(ab')2 (10 ng/ml) for 6 days. Cells were washed and resuspended in media without antigen for 3 days when supernatants were collected. All supernatants were tested for human antibody against 17-1A F(ab')2 and A5C3 F(ab')2 in an ELISA for human IgG. Results are expressed as specific antibody produced per 10^6 B cells cultured on day 0

These data suggest that the initial antigenic stimulus to induce antimurine antibody in these patients is associated with the unique idiotope present on 17-1A but absent on A5C3. Once induced, the polyclonal antibody response contains antibodies with recognized determinants on idiotypically unrelated molecules. This suggests the presence of 17-1A Id-specific T_H cells in these treated patients – an area which is under continuing investigation. In addition, it appears that the predominant specificity of Ig induced by 17-1A F(ab')2 is against isotypic determinants on the molecule.

2.5 Purification and Characterization of Human Anti-Id

The anti-17-1A Id antibody was purified from pooled serum samples from three patients (indicated by the dotted arrows in Fig. 1) by immunoadsorbtion to remove anti-isotypic Ig and to select for 17-1A F(ab')2-binding Ig (Koprowski et al. 1984). The specificity of these purified human anti-Id is shown in Table 4. Binding to 17-1A Mab was significantly higher than to three other Mabs, two of which ($C_4$2032 and $C_4$1472) detect antigens on CRC cells other than the epitopes recognized by 17-1A Mab.

The cross-reactivity between human anti-Ids isolated from sera of three CRC patients was examined in a competition assay for ^{125}I-17-1A. It was found that each human anti-Id cross-competed with another to a significant degree (Koprowski et al. 1984). The degree of inhibition ranged from 40% to 71%. These data demonstrated that although these three unrelated patients produced polyclonal populations of anti-Id, a significant proportion of these molecules bore the same specificity for 17-1A Id.

To insure that the purified human anti-Id was directed against the antigenic site of 17-1A Mab, competition inhibition studies were performed. The binding of 17-1A to CRC cells was assayed in the presence of varying concentrations

Table 4. Binding of purified human anti-Id 17-1A antibodies to monoclonal antibodies

Patient	Anti-Id, μg/ml	Binding of ^{125}I-labeled Mab to anti-Id (cpm)			
		17-1A	$C_4$2032	$C_4$1472	A5C3
08	<5.0	1693±131	214±31	308±44	166±18
07	6	1023±91	168±26	363±21	251±29
23	18	21339±156	427±24	512±59	813±101
	7	7956±293	208±15	NT	427±23

The activity bound to normal human IgG (40 μg/ml) of ^{125}I-labeled 17-1A, $C_4$2032, and A5C3 were 151±25, 139±32, and 247±32, respectively. Results represent mean±SD of triplicate determinations; NT, not tested

Table 5. Inhibition of 17-1A Mab binding to CRC SW116 cells by human anti-Id antibody

Patient	Human anti-Id (μg/ml)[a]	Inhibition of ^{125}I-17-1A binding to SW1116 cells (%)
08	2.0	94.3
	0.2	17.6
	0.02	2.4
47	2.0	30.4
	0.2	0.0
	0.02	0.4

[a] Normal human IgG inhibited ^{125}I-17-1A binding by 16.0%, 0.6%, and 17.8% at the same concentrations

of purified human anti-Id or normal human IgG. The results listed in Table 5 indicate that this human anti-Id successfully inhibited the binding of ^{125}I-17-1A Mab to CRC cell line SW 1116 in a dose-dependent manner while normal human Ig showed no such effect.

3 Induction of Ab3 in GIC Patients

3.1 In Vitro Stimulation with Human Anti-Id (Ab2) Induces Ab3-Like Antibodies

Since Ab2 can be detected in the serum of patients receiving 17-1A Mab and can be induced from 17-1A F(ab′)2-stimulated PBMC in vitro, it was apparent that the Ab2-Id B cells in these patients may have induced the production of Ab3 antibody. This Ab3 antibody would be difficult to detect in serum since, if it exists, it most likely would be bound to Ab2 as immune complexes. Therefore, we attempted to induce Ab3 in vitro from PBMC of two CRC

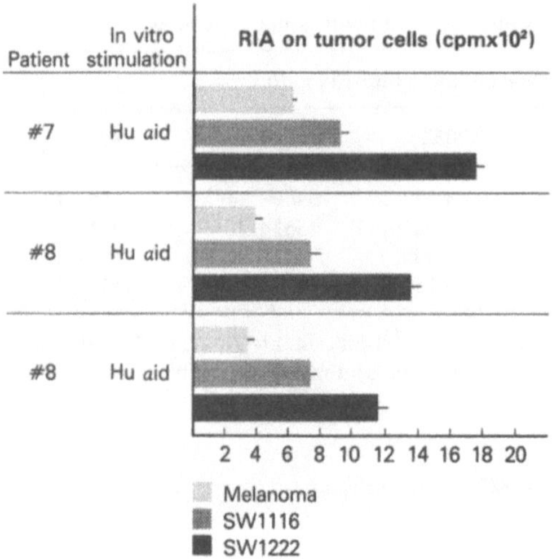

Fig. 3. Induction of Ab3 (anti-anti-Id) which bind to live CRC tumor cells. Experiments were performed as described in Fig. 2 using PBMC from patients 7 and 8 stimulated with autologous anti-Id antibody (Ab2). Supernatants were tested in a live cell RIA using CRC and melanoma ells as targets. Stimulation of PBMC of patient 8 was done in two separate experiments from blood samples obtained 2 weeks apart

patients stimulated with autologous Ab2 purified from their serum. After 6 days, the cells were extensively washed, recultured for 3 days in the absence of Ab2, and supernatants tested for their ability to bind to tumor cells which are 17-1A antigen positive or negative. The results of three experiments are shown in Fig. 3. Both patients had received 17-1A immunotherapy at least 1 year previously and showed transient remission of disease. After stimulation of their PBMC with autologous anti-Id (Ab2), culture supernatants showed positive binding to two CRC 17-A antigen-positive cell lines (SW1116 and SW1222). Less reactivity was seen on a 17-1A antigen-negative cell line (melanoma Carney).

These initial experiments were encouraging in that despite the observation that Ab2β (homobodies) which mimic the antigen defined by Ab1 are present in low frequencies compared with other types of Ab2 (Bona et al. 1984), they can mimic the stimulatory property of the 17-1A antigen for lymphocytes.

3.2 In Vivo Immunization with Human Anti-Id (Ab2)

The ability of purified human Ab2 from several CRC patients to induce Ab3 from both homologous (Fig. 3) and heterologous patients' PBMC (data not shown) suggested that in vivo administration of purified human anti-Id may boost clones of Id-positive lymphocytes primed originally by 17-1A antigen present on the CRC of the patient's own tumor.

For this immunotherapy trial, we selected patient 07, who had Duke D rectal adenocarcinoma resected in July 1981. She had metastases to ischiorectal space and liver. She received 113 mg MAb 17-1A on December 15, 1981 i.v. and 15 mg by local infiltration into the ischiorectal fossa. She responded well to immunotherapy with Mab, since biopsy in June 1982 failed to show presence of liver metastases and no tumor in the abdominal cavity was detected on

Table 6. Specificity of goat IgG specific for the idiotopes of Mab 17-1A and 173-2-35-3

^{125}I-labeled tracer[a]	^{125}I-tracer bound (cpm) catcher antibody	
	goat anti-Id (17-1A)	goat anti-Id (173-2-35-3)
Murine Mab (class)		
17-1A (γ2a)	24478	255
173-2-35-3 (γ2a)	276	10980
A5C3 (γ2a)	207	623
37-7 (γ2a)	564	ND
C420 (γ2a)	342	580
19-9 (γ1)	664	179
Heterologous IgG		
Mouse	249	91
Human	181	148
Rabbit	379	363

[a] Murine Mab and heterologous IgG were labeled with ^{125}I (4–5 μCi/μg protein) and 50 ng incubated with catcher antibody coupled to Sepharose polystyrene beads (Koprowski et al. 1984); ND, not determined

laparotomy in October 1983. This lasted until May 1984, when a mass became visible in the right lobe of the lung. This patient developed antimouse antibodies lasting from 11 through more than 278 days after administration of Mab. Anti-Id antibodies against 17-1A Id were detected in serum of the patient drawn on 18, 124, 614, 656, and 761 days after Mab. Her anti-Id cross-reacted with other anti-Id isolated from two patients who received 17-1A immunotherapy. Two and one-half years after exposure to 17-1A Mab, she received 750 μg purified heterologous human anti-Id intradermally in alum (first injection) and calcium phosphate gel (later injections) on weeks 0, 1, 3 and 6. Peripheral blood for PBMC isolation was collected several hours before each immunization and on week 2, 4, and 5. Mishell-Dutton cultures of her PBMC at each time point were stimulated in vitro with 17-1A F(ab)′2, or the human anti-Id preparation used for the therapy, and supernatants were collected for antibody assays.

In order to detect Ab3 in these supernatants, a monospecific anti-Id antibody was prepared in goats by immunization with 17-1A Mab and selective absorption of the Ig with other murine Mab. The resulting goat anti-Id IgG was specific for 17-1A Mab Id as indicated in Table 6. Of a panel of Mabs of different Id, goat anti-Id (17-1A) bound only to 17-1A Mab and not to other IgG2a mouse Mab. Conversely, a goat anti-Id against 173-2-35-3 Mab (an IgG2a mouse Mab detecting a different CRC antigen) did not detect 17-1A Id but only 173-2-35-3.

During the course of injection of heterologous human anti-Id (Ab2) given patient 07, the production of autologous Ab3 by stimulated PBMC was determined by assaying culture supernatants for binding to goat 17-1A anti-Id (Ab2) and human 17-1A antigen-positive CRC cell lines. In addition, concomitant production of Ab2 against 17-1A Mab was also evaluated. The results are shown in Fig. 4.

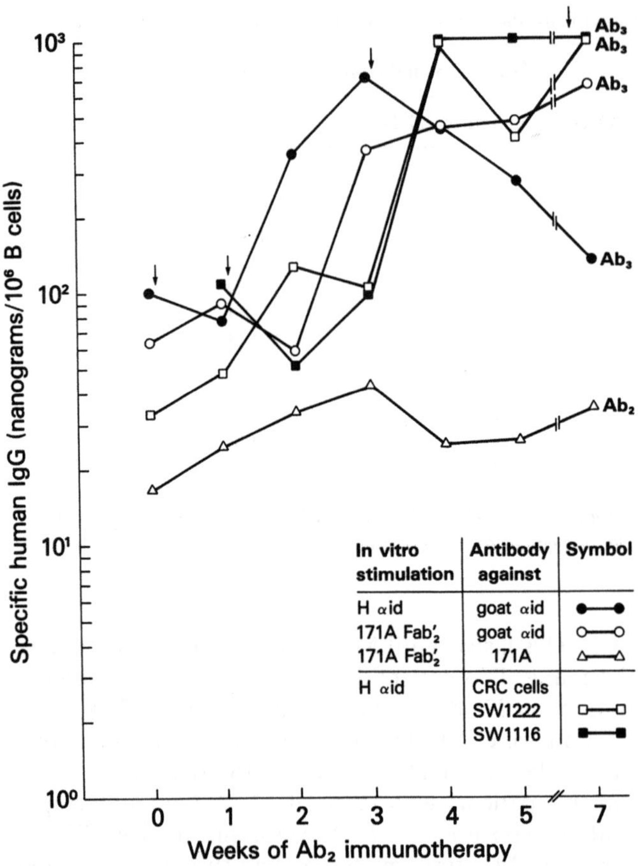

Fig. 4. Induction of Ab2 and Ab3 after in vitro stimulation of PBMC from GIC patient treated with human anti-Id. PBMC taken at various times during therapy were stimulated in vitro with heterologous human anti-Id or 17-1A F(ab′)2. Culture supernatants were tested for Ab2 and Ab3 in an ELISA against 17-1A F(ab′)2 and goat anti-Id respectively. Ab3 which mimics the binding of Ab1 was detected by a live cell RIA against 17-1A antigen-positive CRC tumor cell lines SW 1222 and SW 1116

PBMC obtained *before* immunotherapy were capable of producing Ab3 which bound to goat anti-Id (Ab2) when stimulated in vitro with human anti-Id. Antigenic stimulation with Ab1 [17-1A F(ab′)2] at this time also resulted in increased Ab3 production over controls receiving no antigen.

During the treatment of the patient, Ab3 (as detected by goat anti-Id) peaked by week 3 and declined steadily thereafter. Ab3 binding to CRC cells reached a maximum by week 4 and remained at high levels until after week 7. The specificity of Ab3 for tumor cells induced in this system is being evaluated. Preliminary experiments indicate low but significant reactivity on other tumor types, such as melanoma (see Fig. 3). Whether this reactivity is due to the same population of molecules (Ab3) binding both anti-Id and CRC cells is unknown.

In vitro stimulation with 17-1A F(ab′)2 resulted, in addition to production of Ab3, in the production of Ab2. For this reason, the total amounts of Ab2 and Ab3 in these culture supernatants may be underestimated since a proportion of Ab2 and Ab3 was probably bound in complexes. Simultaneous production of both types of antibodies in response to Ab1 suggests that the network interactions (Ab1 → Ab2 → Ab3) can be recreated in vitro during the 9-day culture period and most probably operate via the recognition and expansion of Id-positive B cells present in the peripheral circulation.

4 Conclusions and Summary

Results presented show that a fraction of the antimurine antibody present in serum of patients injected with mouse Mab represents an anti-Id binding specifically to the Mab injected and not to other Mabs of the same isotype which similarly react with antigens expressed on CRC cells. Presence of anti-Id in serum was further confirmed by its production in tissue culture by lymphocytes of the same patients. Inhibition of the anti-Id binding to Id by extracts of CRC cells and not of other human tumors led one to believe that anti-Id carries the internal image of the tumor antigen. One would expect therefore that presence of anti-Id in patients would elicit formation of Ab3 reacting with anti-Id and with the CRC cells. The Ab3 was indeed produced by lymphocytes of patients in tissue culture following stimulation by homologous anti-Id. Furthermore, immunization of a patient with human anti-Id resulted in an apparent rise of production of Ab3 by lymphocytes stimulated in vitro by anti-Id. The study of this phenomenon is complicated because subjects who were studied are already tumor-bearing and were previously treated with Mab. In order to pursue the study of this phenomenon further, it may be necessary to immunize patients with heterologous anti-Id who were not previously exposed to Mab. The immunization procedure should include various dosages of anti-Id, since it has been shown (in case of rabies anti-Id) (Koprowski et al. 1985) that high concentrations of anti-Id may inhibit formation of Ab3. Following such immunization, attempts will be made to isolate Ab3 either from serum or from stimulated lymphocytes and then to study thoroughly its reaction with tumor cells and normal tissue. In addition to testing reactivity to tumor cells, attempts will be made to show whether Ab3 has a tumoricidal effect either by interaction with human complement or as mediator for effector cells. If this will be the case, a patient's own tumor, if available, could be used as target for the action of Ab3. If successful, these studies may conclusively point to the possible (and desirable) role of Ab3 in control of tumor growth. Late regression of cancer observed in some cases after treatment with Mab (Koprowski et al. 1984; Sears et al. 1984) may then be explained by the engendering of anticancer antibody by the patient himself; this in turn may have contributed to tumor regression. This "indirect form" of immunotherapy seems to be more desirable than any other form inclusive even of "passive" administration of Mab. Hopefully, this indeed will be the case.

References

Bona C, Kohler G (1984) Anti-idiotype antibodies and internal images in monoclonal and anti-idiotype antibodies. In: Probes for receptor structure and function. Liss, New York

Bona C, Victor-Kobrin C, Manheimer A, Bellon B, Rubenstein L (1984) Regulatory arms of the immune network. In: Idiotypic networks. Immunol Reviews, vol 79. Munksgaard, Copenhagen

Cazenave P (1977) Idiotypic-anti-idiotypic regulation of antibody synthesis in rabbits. Proc Natl Acad Sci USA 74:5122

DeFreitas E, Vella S, Linnenbach A, Zmijewski C, Koprowski H, Croce C (1982) Antigen-specific human T cell hybridomas with helper activity. Proc Natl Acad Sci USA 79:6646

Fernandez C, Moller G (1979) Antigen-induced strain-specific auto-anti-idiotypic antibodies modulate the immune response to dextran β512. Proc Natl Acad Sci USA 76:5944

Goidl E, Schrater A, Suskind G, Thorbecke G (1979) Production of auto-anti-idiotypic antibody during the normal immune response to TNP-Ficoll. J Exp Med 150:154

Grossman Z (1982) Recognition of self, balance of growth and competition: horizontal networks regulate immune responsiveness. Eur J Immunol 12:747

Grossman Z, Asofsky R, DeLisi C (1980) The dynamics of antibody secreting cell population: regulation of growth and oscillations in the response to T-independent antigens. J Theor Biol 84:49

Herlyn D, Koprowski H (1982) IgG2a monoclonal antibodies inhibit human tumor growth through interaction with effector cells. Proc Natl Acad Sci USA 79:4761

Herlyn M, Steplewski Z, Herlyn D, Koprowski H (1979) Colorectal carcinoma-specific antigen: detection by means of monoclonal antibodies. Proc Natl Acad Sci USA 76:1438

Herlyn D, Steplewski Z, Herlyn M, Koprowski H (1980) Inhibition of growth of colorectal carcinoma in nude mice by monoclonal antibody. Cancer Res 40:717

Jerne N (1974) Towards a network theory of the immune system. Ann Immunol (Paris) 125C:373

Kennedy R, Melnick J, Dreesman G (1984) Antibody to hepatitis B virus induced by injecting antibodies to the idiotype. Science 223:930

Kennett R, Jonak Z, Bechtol K (1980) Monoclonal antibodies against human tumor-associated antigens. In: Monoclonal antibodies. Plenum, New York

Koprowski H (1984a) Mouse monoclonal antibodies in vivo. In: Monoclonal antibodies and cancer. Academic, New York

Koprowski H (1984b) Human tumor antigens. In: Cancer cells of the transformed phenotype. Cold Spring Harbor, New York

Koprowski H, Steplewski Z (1982) Recognition of human tumor antigens by monoclonal antibodies. In: Genes and tumor antigens. Raven, New York

Koprowski H, Steplewski Z, Mitchell K, Herlyn M, Herlyn D, Fuhrer P (1979) Colorectal carcinoma antigens detected by hybridoma antibodies. Somatic Cell Genet 5:957

Koprowski H, Herlyn D, Lubeck M, DeFreitas E, Sears H (1984) Human anti-idiotypic antibodies in cancer patients: is the modulation of the immune response beneficial for the patient? Proc Natl Acad Sci USA 81:216

Koprowski H, Reagan KJ, Macfarlan RL, Dietzschold D, Wiktor T (1985) A new genertion of rabies vaccines. Rabies glycoprotein gene recombinants, anti-idiotypic antibodies, and synthetic peptides. In: Modern approaches to vaccines. Cold Spring Harbor Symposium, Cold Spring Harbor

Levy R, Miller R (1983) Tumor therapy with monoclonal antibodies. Fed Proc 42:2650

Lindemann J (1979) Homobodies: do they exist? Ann Immunol (Paris) 130C:311

Mach JP, Chatal J, Lumbroso J, Buchegger F, Forni M, Ritschard J, Berche C, Douillard J, Carrel S, Herlyn M, Steplewski Z, Koprowski H (1983) Tumor localization in patients by radiolabeled monoclonal antibodies against colon carcinoma. Cancer Res 43:5593

Miller R, Maloney D, Warnke R, Levy R (1982) Treatment of B-cell lymphoma with monoclonal anti-idiotype antibody. N Engl J Med 306:517

Moldofsky P, Sears H, Mulhern C, Hammond N, Powe J, Gatenby R, Steplewski Z, Koprowski H (1984) Detection of metastatic tumor in normal-sized retroperitoneal lymph nodes by monoclonal antibody imaging. N Engl J Med 311:106

Powe J, Herlyn D, Alavi A, Munz D, Steplewski Z, Koprowski H (1985) Radioimmunodetection

of human tumor xenografts by monoclonal antibodies correlates with antibody density and affinity. Proc Eur Assoc Nucleic Med (to be published)
Reagan K, Wunner W, Wiktor T, Koprowski H (1983) Anti-idiotypic antibodies induce neutralizing antibodies to rabies virus glycoprotein. J Virol 48:660
Ritz J, Pesando J, Sallan S, Clavell L, Notis-McConarty J, Rosenthal P, Schlossman S (1981) Serotherapy of acute lymphoblastic leukemia with monoclonal antibody. Blood 58:141
Saxon A, Feldhaus J, Robbins R (1982) Single step separation of human T and B lymphocytes using AET-treated SRBC rosettes. J Immunol Methods 12:285
Sears H, Herlyn D, Steplewski Z, Koprowski H (1984) Effects of monoclonal antibody immunotherapy on patients with gastrointestinal adenocarcinoma. J Biol Response Mod 3:138
Sege K, Peterson P (1978) Use of anti-idiotypic antibodies as cell-surface receptor probes. Proc Natl Acad Sci USA 75:2443
Steplewski Z, Lubeck M, Koprowski H (1983) Human macrophages armed with murine immunoglobulin G2a antibodies to tumors destroy cancer cells. Science 221:865
Urbain J, Wikler M, Franssen JD, Callignon C (1977) Idiotypic regulation of the immune system by the induction of antibodies and anti-idiotypic antibodies. Proc Natl Acad Sci USA 74:5126
Wands J, Marciniak R, Isselbacher K, Varghese M, Don G, Halliday J, Powell L (1982) Demonstration of previously undetected hepatitis B viral determinants in an Australian Aboriginal population by monoclonal anti-hepatitis antibody radioimmunoassays. Lancet 1:977

Recognition of Physiological Receptors by Anti-Idiotypic Antibodies: Molecular Mimicry of the Ligand or Cross-Reactivity?*

A.D. STROSBERG, J.G. GUILLET, S. CHAMAT, and J. HOEBEKE

* This work was supported by grants from the *Centre National de la Recherche Scientifique, Institut National de la Santé et de la Recherche Médicale (contrat "Recherche Externe"), Association pour le Développement de la Recherche sur le Cancer, Fondation de la Recherche Médicale Française, Ligue de la Recherche contre la Cancer,* University of Paris VII, and *Ministère de l'Industrie et de la Recherche.*

Laboratory of Molecular Immunology, CNRS, Institut Jacques Monod, University Paris VII, 2, place Jussieu, F-75251 Paris Cedex 05

Current Topics in Microbiology and Immunology, Vol. 119

1 Introduction

The specific binding and physiological modulation of physiological membrane receptors by anti-idiotypic (anti-Id) antibodies has now been described in several systems (reviewed in STROSBERG and SCHREIBER 1984). Two types of explanations have been given for these unexpected interactions. The anti-Id antibody either acts as an internal image of the ligand recognized by both the physiological receptor and the anti-ligand antibody (JERNE 1974, 1984) or, alternatively, there is recognition, by the anti-Id antibodies, of structural homologies shared by the Id-bearing antiligand antibodies and the receptor (STROSBERG 1984). We have represented these two alternatives in Fig. 1 where Ab1 represents the mono-

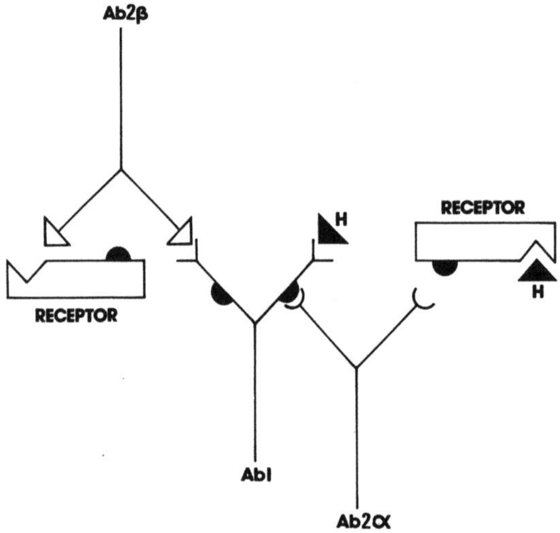

Fig. 1. Immunological interactions involving a hormone and its specific receptor and antibodies

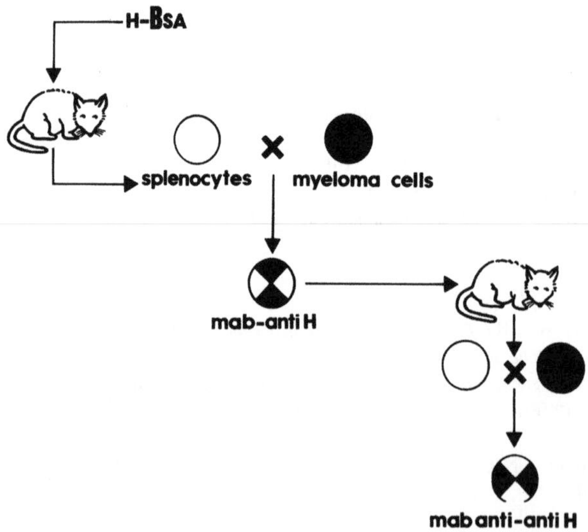

Fig. 2. Preparation of monoclonal antihormone and anti-antihormone antibodies

clonal antihormone antibody and Ab2α and Ab2β stand for the cross-reactive or the internal image anti-Id antibodies, respectively. To choose between these two possibilities, we have begun studies using monoclonal rather than the previously used polyclonal antibodies. Based on our own and others' experience we devised a number of improved assays which will be described here together with the results obtained.

The various steps of the procedure are described, and some of the possible difficulties in interpreting the results are discussed.

The first step is selection of a monoclonal anti-ligand antibody, Ab1, which exhibits binding properties to the ligand most similar to those of the receptor. Then, monoclonal anti-Id antibodies, Ab2, are raised against it (Fig. 2). Finally, the interactions of the anti-Id antibodies with the physiological receptor are studied. A similar approach was recently described for the production of monoclonal anti-Id antibodies which recognize the mammalian reovirus receptors (NOSEWORTHY et al. 1983; KAUFFMAN et al. 1983).

2 The Anti-Ligand Antibody Ab1

2.1 Methodological Considerations

2.1.1 Choice of the Ligand

The most appropriate ligand molecules to raise antibodies that will allow the induction of a receptor-specific anti-Id response should be as small and as rigid as possible. The larger the ligand, the higher the probability that antigenic determinants will be recognized which are irrelevant to the binding of the ligand to the receptor. Hormones will induce polyclonal antibodies, which in part will bind to the determinant recognized by the receptor. However, since monoclonal antibodies are monospecific, the binding of the ligand to the individual monoclonal antibody must be tested with smaller fragments of the antigen known to be involved in receptor recognition.

The coupling of the ligand to the carrier molecule could result in a large decrease in the affinity of the receptor for the ligand. This may indicate that parts essential for binding are lost by the coupling. Thus, the antibodies induced by this ligand-carrier molecule may only poorly reflect the binding properties of the receptors.

For the study of the immunogenicity of β-adrenergic ligands, we chose the antagonist alprenolol, since its reactive vinyl moiety on the phenyl ring may be modified without loss in affinity for the β-adrenergic receptor (HOEBEKE et al. 1978).

2.1.2 Choice of the Carrier

To raise anti-hapten antibodies, a sufficient amount of substitution of the carrier molecule is needed. To increase the number of reactive residues it may be advantageous to unfold the carrier protein in 6 M guanidinium chloride and to reduce all existing disulfide bonds before coupling of the ligand (HOEBEKE et al. 1978). One may also modify the native protein with multifunctional reagents, e.g., succinylation of the amino groups to increase the number of carboxylic groups capable of reacting, after carbodiimide treatment, with primary or secondary amino groups of the ligand (KAMEL et al. 1979).

2.1.3 Characterization of the Monoclonal Antibodies

A first screening of hybridoma supernatants is most easily performed by an enzyme immunoassay using the ligand coupled to a carrier protein different from that used for immunization (CHAMAT et al. 1984). When highly radioactive ligands for the receptor are available, the affinity of the antibodies present in the hybridoma supernatants can be determined by a radioimmunoassay (FARR 1958). The choice of a monoclonal antibody with properties similar to the receptor can further be verified in an inhibition enzyme immunoassay using, as competitors, molecules which are structurally as little as possible related to the hapten but with a sizable affinity for the receptor (HOEBEKE et al. 1983). Molecules, similar to the hapten but which are not specific for the receptor, should also be tested for their affinity for the antibody.

2.2 Catecholamine-Binding Proteins

2.1.1 The β-Adrenergic Receptor-Cyclase Complex

Catecholamine hormones exert their effects on various tissues through α and β adrenergic receptors. Agonists such as adrenaline or isoproterenol stimulate adenylate cyclase through β_1 and β_2 adrenergic receptors and this effect is inhibited by antagonists such as propranolol and alprenolol (LEFKOWITZ et al. 1983). The receptors have been purified using affinity chromatography and specific antibodies have been prepared against them. These antibodies mimic β-adrenergic ligands by binding to the receptors and by stimulating cyclase activity. The antibodies identify the β receptors by Western blotting and may be used for immunoprecipitation and immunovisualization experiments (COURAUD et al. 1983; STROSBERG et al. 1982).

2.2.2 Antibodies Against Catecholamines

The difficulties in preparing sufficient quantities of β-adrenergic receptors have hampered the large-scale production of both polyclonal and monoclonal antibodies. This led us to explore alternative ways. Antibodies were prepared against alprenolol for the purpose of inducing anti-Id antibodies.

2.2.2.1 *Binding Properties of Anti-Alprenolol Antibodies*

Anti-alprenolol antibodies were obtained in rabbits (HOEBEKE et al. 1978) and in mice (CHAMAT et al. 1984) after immunization with alprenolol coupled to bovine serum albumin. The apparent affinity constant of the serum antibodies for free alprenolol was respectively 1.4×10^8 M^{-1} and 0.2×10^8 M^1. Both the rabbit and the mouse antibodies also recognized other β-adrenergic ligands, with a higher affinity for antagonists than for agonists.

2.2.2.2 *Binding Properties of Monoclonal Anti-Alprenolol Antibodies*

Four monoclonal anti-alprenolol antibodies were prepared by somatic cell fusion between immune murine splenocytes and myeloma NS1 cells. The affinity of one of these antibodies, 37A4, for alprenolol was ten times higher (10^8 M^{-1}) than that of 37A11, 10E2 and 22C4 (10^7-10^6 M^{-1}).

Two of the antibodies, 37A11 and 22C4, did not show any stereospecificity for either agonist or antagonist binding. Antibody 10E2 was stereospecific for both agonists and antagonists while 37A4 was only stereospecific for antagonists. This stereospecificity was similar to that of the receptor, the l-isomer being better recognized than the d-isomer. Further characterization of the binding with several β-adrenergic agonists and antagonists showed that 37A4 was specific for β antagonists irrespective of the β_1 or the β_2 specificity, and that all the β agonists, including the natural hormones, adrenaline and noradrenaline, were recognized. In contrast, the 10E2 antibody did not recognize the hormones while recognizing the antagonists (Table 1). Finally, 37A4 showed an increased intrinsic fluorescence upon alprenolol binding, suggesting the formation of a hydrophobic complex between a tryptophan residue and the vinyl-phenyl ring of alprenolol similar to that found on the β_2-receptor of frog erythrocytes (CHERKSEY et al. 1981).

Subsequent fusions yielded 12 other monoclonal anti-alprenolol antibodies which were characterized for their isotype, isoelectric point, and affinity constant for alprenolol. The association constants for alprenolol ranged from 2 to 200×10^6 M^{-1}.

2.2.2.3 *Structural Studies*

Whole antibodies and separated heavy and light chains of four monoclonal antibodies were submitted to direct automated Edman degradation. Despite several attempts, only the K light chain of antibody 22C4 yielded an amino terminal sequence. Thus, the three K chains from 37A4, 37A11, and 10E2 appeared to be blocked, possibly as the consequence of a cyclized glutamine residue. A survey of the literature indicates that blocked immunoglobulin chains are extremely rare in the mouse. The V_H region of 37A4 could be directly sequenced over the first 25 residues. This was not possible for the H chains of the three other antibodies which again appeared to be unavailable to direct Edman degradation.

Table 1. Binding of different adrenergic ligands to the 37A4 and the 10E2 monoclonal antibodies

Compounds	Receptor subtypes				Affinity constants (K_A or K_I)	
	α_1	α_2	β_1	β_2	37A4	10E2
Antagonists					$10^6\ M^{-1}$	
Alprenolol[a]	−	−	+	+	24.0	6.6
CGP-12,177[a]	−	−	+	+	10.3	2.4
Propranolol[b]	−	−	+	+	15.0	4.2
Practolol[b]	−	−	+	−	0.11	0.068
Yohimbine[b]	−	+	−	−	No detectable binding	
Prazosin[a]	+	−	−	−	No detectable binding	
Agonists					$10^3\ M^{-1}$	
Adrenaline	+	+	+	+	1.6	No detectable binding
Noradrenaline	+	+	+	+	1.0	No detectable binding
Isoproterenol	−	−	+	+	2.0	13.3
Salbutamol	−	−	+	+	7.9	0.5
Procaterol	−	−	−	+	1.0	No detectable binding

[a] Association constants were determined by direct saturation binding experiments using tritiated compounds

[b] Inhibition constants were calculated from inhibition curves using the method of CHENG and PRUSSOFF. The inhibition constants for all agonists were calculated in the same way

The DNA of the hybridoma cells was fragmented by different restriction enzymes and Southern blots of the digests were analyzed with the help of probes specific for the J_K and J_H loci. Preliminary results indicate that 37A4, 37A11, and 10E2 use the same J_K and V_K genes. This was not the case for the V_K gene expressed in the 22C4 line. For the heavy chains, similar experiments suggest that 37A11 and 10E2 express the same germline V_H gene, whereas 37A4 uses a different one. Thus, the results obtained by Southern blotting analysis are in agreement with the amino acid sequence data and suggest that H and L chains from anti-alprenolol antibodies appear to use particular subgroups infrequently found in other antibodies.

2.3 Hormone-Binding Antibodies (Ab1) Involved in Other Systems

2.3.1 Antibodies Against a Potent Agonist of the Nicotinic Acetylcholine Receptor

WASSERMAN et al. (1982) used a derivative of BIS-Q (trans −3,3′ bis-a-trimethylammonio) methylazobenzene bromide, a potent agonist of the nicotinic acetylcholine receptor, to prepare rabbit antibodies which mimicked the binding characteristics of the acetylcholine receptor. They suggested that since agonists bind to acetylcholine receptor when it is in its activated state only, antisera are homologous to an activated rather than a resting state receptor. Antibodies against an agonist (i.e., BIS-Q) are able to recognize a related agonist (decamethonium) 500 times better than the antagonist hexamethonium.

CLEVELAND et al. (1983) immunized mice with a bovine serum albumin conjugate of BIS-Q and were able to expand a population of spleen cells that secreted antibodies specific for BIS-Q (Ab1). The fact that they also obtained from the same mice a significant anti-anti-BIS-Q (Ab2) response against nicotinic acetylcholine receptors supports the hypothesis of possible similarities between antibody and receptor.

2.3.2 Antibodies Against Opioid Peptides or Against Morphine

Antibodies against human β-endorphin were prepared in mice. Resulting monoclonal antibodies displayed different binding patterns with the antigen (MEO et al. 1983). By cross-reactivity studies, the authors defined the specific sequence Tyr-Gly-Gly-Phe as the antigenic determinant recognized by one of the monoclonal antibodies (Table 2).

The tetrapeptide Tyr-Gly-Gly-Phe represents the "message" sequence found at the NH_2 terminus of all naturally occurring mammalian opioid peptides. Using histochemical analysis, the same authors showed that this monoclonal antibody labeled neurons containing opioid peptides. The antigenic determinant recognized by this antibody may also contain the information necessary for specific recognition by mammalian opioid receptors. Recently, NG and ISOM (1984) obtained antimorphine antibodies in rabbits. The polyclonal antibodies showed selective binding to various opiate ligands and reacted differentially in a stereospecific manner.

Unlike the opiate receptor, antimorphine antibodies bound agonists with a higher affinity than antagonists. However the hypothesis proposed by Wasserman could be used to explain the results obtained by Ng: the antimorphine antibodies described could also mimic the opiate receptor in an activated state.

2.3.3 Antibodies Against the Formyl Chemotactic Peptide

MARASCO et al. (1982) produced antibodies of restricted heterogeneity against the chemotactic peptide fMet-Leu-Phe.

Analysis of the structural requirements for recognition of this peptide by the antibody showed great similarities with those for recognition by the receptor. Both recognitions require the presence of the formyl group on the methionine residue of the hydrophobic leucine, and of the hydrophobic benzyl side chain of phenyl alanine. The only marked difference between the antibody and the receptor was the absolute requirement of the carboxyl group of phenylalanine for recognition by the latter, while the former also recognized C terminal-substituted peptides. The reason for this difference according to the authors may reside in the linkage of this group to the carrier protein used for immunization.

Table 2. Cross-reactivities of monoclonal antibody 3-E7

Peptide	Relative activity	Sequence
		1 60 61 69 91
β_h-LPH	<0.1	–//–Arg-Tyr-Gly-Gly-Phe-Met-Thr-Ser-Glu-Lys-Ser-Gln-Thr-Pro–//–Tyr-Lys-Lys-Gly-Glu
β_h-EP	100	————————————————//————————
β_c-EP	100	————————————————// –His————Gln
β_h-EP$_{61-69}$	100	——————————
[Met]Enkephalin	80	—————
[Leu]Enkephalin	80	—————Leu
β_h-LPH$_{60-65}$	90	——————
des-Leu, Met-Enkephalin	5	————
Acetyl-β_h-EP	<0.1	MeCo—————————————————————
des-Tyr$_{61}$-β-EP	<0.1	————————————————————
β_h-LPH$_{66-91}$	<0.1	————————————————
Dynorphin	16	—————Leu-Arg-Arg-Ile-Arg-Pro-Lys-Leu-Lys
α-neo-EP	20	—————Leu-Arg-Lys-Tyr-Pro-Lys

Immunoreactivities are expressed relative to the 50% inhibitory concentration of the homologous antigen β_h-EP (10 pmol per tube)

3 The Anti-Antiligand Antibody Ab2

3.1 The Anti-Idiotypic Anti-β Receptor Response in Rabbits

Since the first anti-alprenolol antibodies were raised in rabbits, the same species was chosen to raise anti-Id antibodies (SCHREIBER et al. 1980). Among the first three animals immunized with a rabbit Ig fraction containing polyclonal Ab1 anti-alprenolol antibodies, only one was found to produce Ab2 anti-Id antibodies which completely inhibited the binding of alprenolol to Ab1. Partial noncompetitive inhibition of the binding to turkey erythrocyte β-adrenergic receptors was also observed.

Several experiments were performed to demonstrate direct binding of Ab2 to β-receptors. This could be visualized through indirect immunofluorescence binding to turkey erythrocytes, human Hela cells, and murine mastocytoma P815 cells, all known to bear such receptors. The binding of Ab2 to the erythrocytes could be quantitated by using radioiodination and the number of bound Ab2 molecules was roughly the same as the number of receptors per cell (700–900 per erythrocyte) suggesting a one-to-one stoechiometry. The binding of Ab2 also modulated the basal and catecholamine sensitive adenylate cyclase activity by increasing it three- to fourfold in our study (SCHREIBER et al. 1980), while this activity was actually inhibited by the anti-Id antibody described by HOMCY et al. (1982).

Inhibition of alprenolol binding to β-adrenergic receptors was used to follow the Ab2 response during the weeks after immunization with Ab1. Although individual variations were observed, the Ab2 anti-β receptor response appeared to be short-lived and characterized by sharp peaks of activity. The cyclical reappearance of anti-β receptor Ab2 was seen in some animals, whereas in others, such antibodies did not appear more than once (COURAUD et al. 1983; LU et al. 1984).

To investigate this unusual and drastic variation, the Ab1-like activity of the sera was tested for the possibility that such antibodies were responsible for the neutralization of Ab2. The Ab1-like activity was found to appear in sharp peaks corresponding to bleedings in which no Ab2 antireceptor activity could be detected.

The Ab1-like activity was compared with that of the injected Ab1 anti-alprenolol antibodies. A detailed analysis revealed that the intrinsic Ab1-like activity was really different from the immunogen both in terms of affinity for the ligand and in the capacity to neutralize Ab2. Thus, it was concluded that the endogenous Ab1-like activity really revealed an "anti-Ab2" response. This hypothesis was supported by the finding that the various Ab2 peaks of neutralizing activity each corresponded to molecules with decreasing binding activity for alprenolol (COURAUD et al. 1983).

To gain further insight into the nature of Ab2 β-adrenergic receptor interactions, it was decided to prepare polyclonal and monoclonal (m) Ab2, directed against m Ab1. A rabbit was immunized with mAb 37A4 and the anti-Id antibodies were characterized for their ability to inhibit the binding of alprenolol to 37A4 and to other anti-alprenolol mAb. While close to 100% inhibition was observed for the homologous reaction, inhibition did not exceed 20% for binding to 37A11 or 10E2. We then proceeded to raise monoclonal Ab2 in mice.

3.2 The Anti-Idiotypic Response in Mice

3.2.1 Polyclonal Anti-Idiotypic Response

Before preparing monoclonal Ab2, we first studied the polyclonal response. Two groups of five mice were immunized each with a different monoclonal anti-alprenolol antibody. In Fig. 3A we show that the response to Ab1 37A4 did not drastically change in time: a single peak was observed which slowly decreased in the 150 days following immunization. In contrast, in Fig. 3B, the Ab2 response to 10E2 appeared to be more similar to that observed in rabbits, i.e., in two of four mice, a sharp peak of Ab1-like activity appeared long after immunization.

3.2.2 The Monoclonal Anti-Idiotypic Antibodies

3.2.2.1 The Immunization Route

One of the advantages of using a murine monoclonal antibody to raise murine monoclonal anti-Id antibodies resides in the fact that the response in a syngeneic organism is limited to the anti-Id response. No anti-isotypic or anti-allotypic antibodies are produced. The injection of the Id antibodies in complete Freund's adjuvant (CFA) could, however, still generate a heterogeneous response since *Mycobacterium butyricum,* the inflammatory agent of the adjuvant is a polyclonal mitogen which may stimulate B lymphocytes unrelated to the anti-Id response. The polyclonal reaction may also involve the production of anticarbohydrate antibodies, reacting with membrane glycoproteins or rheumatoid factors, reacting with the Fc part of antibodies (Bona et al. 1984). To avoid the possibility of such responses, no adjuvant was used, and we injected i.v. paraformaldehyde-fixed 37A4 hybridoma cells, which carry the antibody at their surface (Guillet et al. 1984). A similar immunization method was recently used for the induction of mouse monoclonal anti-Id antibodies against human B lymphomas (Thielemans et al. 1984).

3.2.2.2 Screening for Monoclonal Anti-Idiotype Antibodies

Working in a syngeneic system renders screening more difficult since most assays are based on recognition by a second antibody (radiolabeled or enzyme coupled) which cannot distinguish between the Id and the anti-Id antibodies. We describe here some ways to circumvent that difficulty.

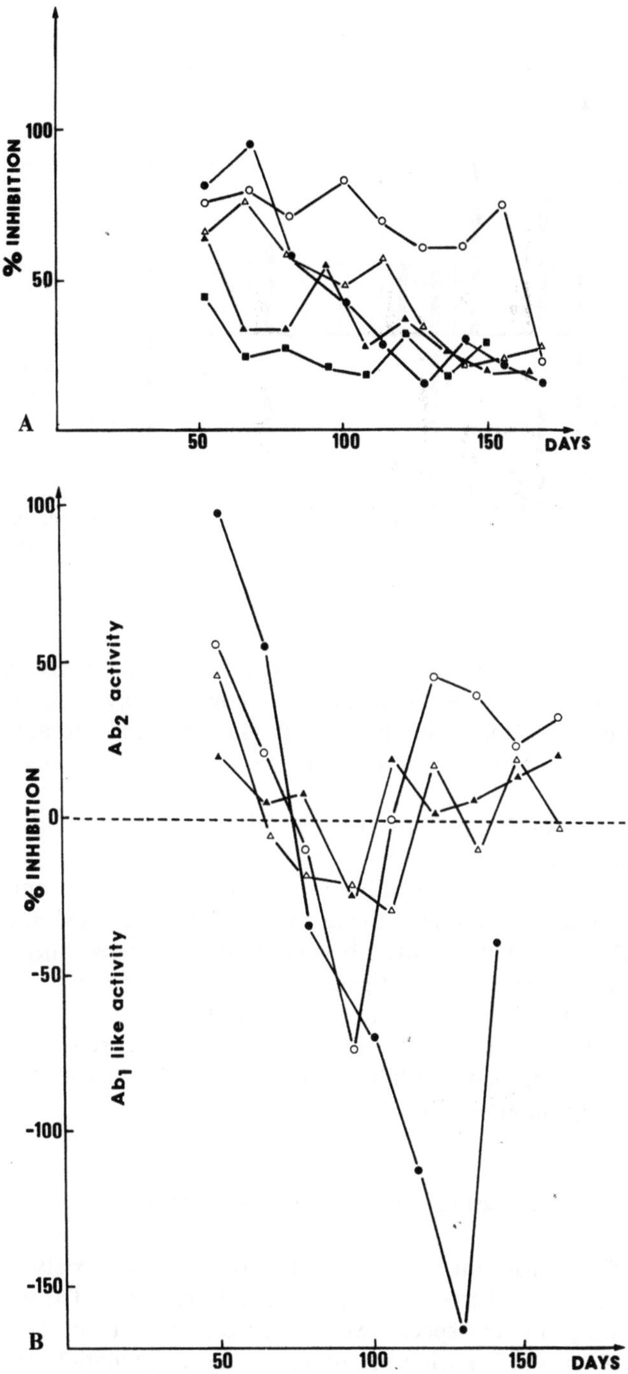

Fig. 3A. Anti-idiotypic activity of mice sera immunized with 37A4 monoclonal antibody. The anti-idiotypic activity was determined by the inhibition of alprenolol binding to the monoclonal antibody. Time curves of five mice are shown (Strosberg et al. 1985). **B** Anti-idiotypic activity of mice sera immunized with 10E2 monoclonal antibody. The anti-idiotypic activity was determined by the inhibition of alprenolol binding to the monoclonal antibody. Negative values correspond to alprenolol-binding activity in the sera. Time curves of four mice are shown (Strosberg et al. 1985)

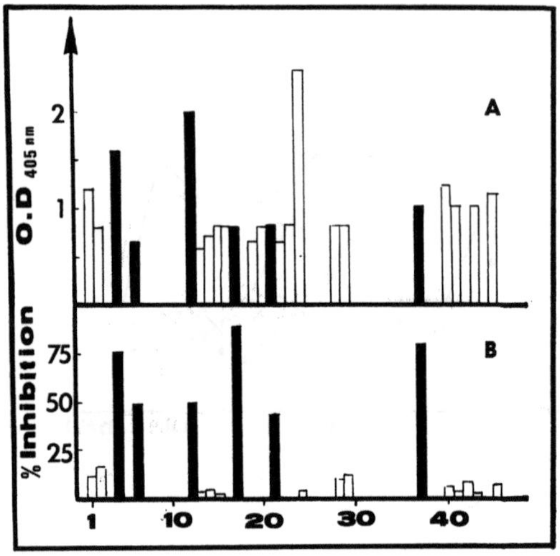

Fig. 4. Screening of hybridomas secreting anti-idiotypic antibodies. **A** Fixation of secretion products on the idiotope bearing 37A4 hybridoma. **B** Inhibition of 37A4 binding on KLH-Alp by the secretion products. The *black blocks* indicate hybridomas secreting antibodies positive in both tests (GUILLET et al. 1984)

Adsorption to the solid phase of the antibody-bearing fixed hybridoma cells instead of the isolated antibody considerably reduces the recognition of the Fc part of the antibody by antibodies, probably because of the low accessibility of the Fc regions at or near the cell surface. Using the appropriate anti-Fc antibody dilution, we can titrate out the response towards the hybridoma cells and thus limit it to the anti-Id bound to the Id-bearing hybridoma cells (Fig. 4A).

Another way to differentiate between the two antibodies (Id and anti-Id) is to radiolabel one of the two types of molecules (MORAHAN 1983). We preferred to use biosynthetically labeled hybridoma products because radioiodination may induce changes in antigen recognition (MASON and WILLIAMS 1980). Intrinsic labeling was also used to perform an Id inhibition radioimmunoassay; as shown in Fig. 4B, six supernatants also passed the more stringent test of anti-Id activity consisting of the inhibition of the interaction between the Ab1 antibody and its ligand alprenolol.

3.2.2.3 Screening for Receptor Recognition

The anti-Id antibodies were tested against a variety of plasma membranes or whole cells carrying the β-adrenergic receptor. Three different cell types from three different species were used: turkey erythrocyte membranes (ATLAS et al. 1974), P815 murine mastocytoma cells (DURIEU-TRAUTMANN et al. 1985), and the human epidermoid cells A431 (DELAVIER-KLUTCHKO et al. 1984). A rabbit B-lymphoma cell, devoid of β receptors, was used as a negative control.

Twenty-four hybridomas recognized the anti-alprenolol antibody 37A4, six inhibited homologous hapten binding, and three of these recognized β-adrenergic receptors.

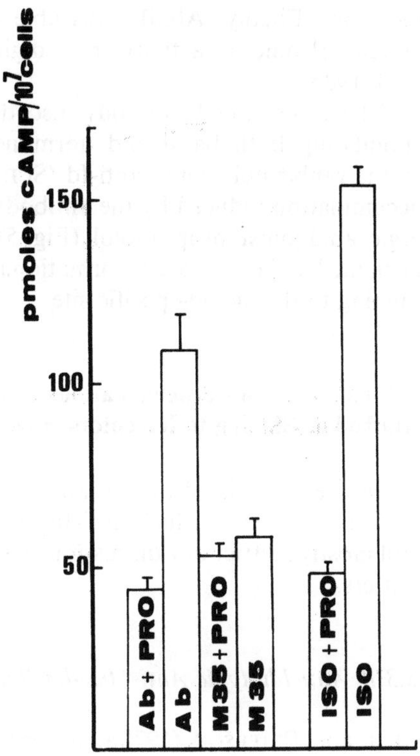

Fig. 5. Stimulation of adenylate cyclase activity of A431 cells by the anti-idiotype Ab2B4. Anti-idiotypic Ab2B4 or M35 (control) antibody purified on an ACA 22 column or the agonist L-isoproterenol were added to the incubation medium with or without the antagonist, L-propranolol; cAMP level was determined on an aliquot of the supernatants by a radiochemical assay using the cAMP kit from Amersham (GUILLET et al. 1985)

3.2.2.4 Detailed Analysis of the Interactions Between a Monoclonal Anti-Idiotypic Antibody and β-Arenergic Receptors

One of the β-adrenergic receptors, the IgM secreting Ab2B4, was further characterized. The anti-Id monoclonal antibody specifically recognized A431 cells known to possess a large number of β_2-adrenergic receptors (DELAVIER-KLUTCHKO et al. 1984). Specific recognition of the β-adrenergic receptors of A431 cells by Ab2B4 was further assessed using immunofluorescence assays. On cells grown near confluence, a characteristic staining of the membranes was detected using Ab2B4, a specific rabbit antimouse IgM, and a fluoresceinated goat antirabbit antibody. When the β-receptor bearing cells were first treated with β-adrenergic agonists, the number of ligand-binding sites sharply decreased through a "desensitization" phenomenon, usually explained by receptor disappearance from the cell surface. Concomitantly, indirect immunofluorescence using Ab2B4 was greatly reduced (GUILLET et al. 1985). These results showed that there is at least one cross-reactive "idiotope-epitope" on anti-alprenolol antibody and β receptors. Further evidence for the recognition of β-receptors by Ab2B4 has come from the positive identification in solubilized A431 plasma membranes, through the technique of "Western blotting", of a polypeptide chain with a mol. wt. of 55 kD, characteristic for the β_2-adrenergic receptors. Finally, Ab2B4 was able to immunoprecipitate up to 80% of the alprenolol binding activity from digitonin treated A431 membranes (GUILLET et al. 1985).

The monoclonal antibody also displayed β-adrenergic agonist activity by stimulating both basal and hormone-sensitive adenylate cyclase. In contrast to the rabbit polyclonal anti-Id (SCHREIBER et al. 1980), the increase in cAMP accumulation induced by the antibody could be totally reversed by the β-adrenergic antagonist propranolol (Fig. 5). The effect of anti-Id Ab2B4 could be explained either by a conformational change of the β receptors or by direct binding to the ligand-specific site.

3.3 Comparison Between Various Systems in which Anti-Idiotypic Antibodies Binding to Receptors Have Been Described

While we have listed all the known observations in Table 3 in comparing the various systems in which anti-Id antibodies act as antireceptor antibodies, we will mostly restrict our discussion to systems involving hormones or neurotransmitters.

3.3.1 Anti-Idiotypic Anti-(Insulin Receptor) Antibodies

SEGE and PETERSON (1978a, b) were the first to show that the injection into rabbits of anti-insulin antibodies resulted in the synthesis of anti-Id antibodies which mimicked the action of the hormone by binding to the insulin receptors on rat epididymal fat cells and by stimulating the uptake by rat thymocytes of aminoisobutyric acid. These observations were confirmed and extended by SHECHTER et al. (1982), who showed that the simple injection of insulin into mice provoked the production of not only anti-insulin antibodies but also of autologous anti-idiotypic antibodies, which competed with insulin both for binding to the receptor and for stimulation of glucose uptake by insulin-sensitive cells. The binding of the anti-idiotypes to the receptor could be blocked by anti-insulin antibodies, presumably through the idiotype-antiidiotype interaction. These studies were all done with polyclonal antibodies, on insulin receptor-bearing cells. The physiological effects considered remain little understood and the precise interactions between the insulin receptor and the effector system are not known.

3.3.2 Anti-Idiotypic Anti-(Nicotinic Acetylcholine Receptor) Antibodies

Immunization of rabbits with the purified anti-(Bis-Q) antibodies preparated by Wasserman et al. (see Sect. 3.3.1) yielded anti-Id sera which recognized rat, *Torpedo*, or eel acetylcholine receptor by complement fixation and enzyme immunoassay. The binding was inhibited by the free-ligand Bis-Q. Two of the three rabbits showed signs of muscle weakness similar to that seen after immuni-

Table 3. Anti-idiotypic antibodies which recognize receptors

Immunogen	Receptor	Antibody-induced signal	Cell	References
Insulin	Insulin receptor	Oxidation and uptake of glucose; Inhibition lipolysis	Adipocyte	SEGE and PETERSON (1978); SHECHTER et al. (1982)
Retinol	RBP	Retinol uptake	Intestinal epithelial cell	SEGE and PETERSON (1978)
Alprenolol	β-adrenergic receptor	Adenylate cyclase activation or inhibition activation	Nucleated erythrocyte	SCHREIBER et al. (1980); HOMCY et al. (1982); GUILLET et al. (1985)
Thyrotropin	Thyrotropin receptor	Adenylate cyclase activation; Increased I transport; Follicle formation	Thyroid cell	FARID et al. (1982)
Chemotactic peptide	Chemotactic peptide receptor		Neutrophil cell	MARASCO and BECKER (1982)
Nicotinic agonist Bis-Q	Nicotinic acetylcholine receptor			WASSERMAN et al. (1982); CLEVELAND et al. (1983)
Morphine	Opiate receptor	Inhibition of electrically induced contractions	Mouse brain	NG and ISOM (1984)
Opioid peptide	Opiate receptor	Inhibition of cAMP synthesis	NG 108CC15	SCHULZ and GRAMSCH (1984)
Reovirus	Receptor for mammalian reovirus	Blocking of viral binding to cells	Lymphoid cells	NOSEWORTHY et al. (1983); KAUFFMAN et al. (1983)

zation with the purified receptor. One of the rabbits was injected intramuscularly with neostigmine and showed temporary improvement. Another showed post-tetanic exhaustion of hindlimb muscles after stimulation of the sciatic nerve. The third rabbit, which had a significant titer of anti-Id antibodies, showed no signs of muscle weakness.

As was the case for the anti-Id response to anti-(β-adrenergic) ligands, the response to anti-(Bis-Q) was transient in one animal, both with respect to antireceptor titer and to signs of experimental myasthenia gravis. Maximal titer and muscle weakness occurred after the first boost and remained high until the second boost, after which the titer dropped three- to fourfold and signs of weakness disappeared. Subsequent boosting apparently caused tolerance: the antireceptor titer dropped to the range found in nonimmunized animals.

When using somatic cell fusion to expand a population of spleen cells secreting anti-Bis-Q antibodies, CLEVELAND et al. (1983) isolated one monoclonal antibody which was found to be directed against other monoclonal anti-Bis-Q antibodies. This spontaneously occurring anti-idiotypic antibody was found to bind to the nicotinic acetylcholine receptor. Its binding to anti-Bis-Q antibodies was found to be stoechiometrically inhibited by purified acetylcholine receptors.

3.3.3 Anti-Idiotypic Anti-(TSH Receptor) Antibodies

Rabbits immunized with rat anti-(human thyroid-stimulating hormone) antibodies produced immunoglobulins which did not bind the hormone but inhibited in a dose-dependent manner up to 50% of the total binding of bovine TSH to the porcine thyroid receptor. Direct interaction of the anti-Id antibodies to porcine thyroid cell membranes was saturable and could be inhibited up to 64% by increasing concentrations of the free bovine hormone. In the presence of Gpp(NH)p (guanosine 5′-(β.-imido)triphosphate) the anti-Id antibodies increased thyroid membrane adenylate cyclase activity 40% over the enzymic activity induced by nonspecific immunoglobulins. The rate of incorporation of (^{131}-I)-labeled sodium iodide into cultured thyrocytes was also increased in a dose-dependent manner by the anti-Id antibodies which induced the organization of these cells into follicles between 5 and 7 days of culture (FARID et al. 1982).

3.3.4 Anti-Idiotypic Anti-Chemotactic Peptide Receptor

MARASCO and BECKER (1982) raised anti-Id antibodies in mice, guinea pigs, and goats against the rabbit antibodies to the chemo-attractant peptide fMet-Leu-Phe (mentioned in Sect. 3.3.3). The goat antibodies bound to rabbit polymorphonuclear leukocytes and the corresponding $F(ab')_2$ fragments partially inhibited the binding of the fMet-Leu-Phe peptide to the same cells. Both the anti-(polymorphonuclear receptor) and the anti-Id activities were lost after passage over an affinity gel containing antibody against the fMet-Leu-Phe peptide. The authors were unable to demonstrate that the anti-Id antibodies mimicked the biological activity of this peptide, but this was explained by the fact that the preimmune control IgG preparations themselves induced locomotion and, in the presence or absence of cytochalasin B, granule enzyme release. These "nonspecific" effects thus could mask a hypothetical specific effect by the anti-Id "internal image".

3.3.5 Anti-Idiotypic Antibodies to Opiate Receptors

SCHULZ and GRAMSCH (1984) obtained anti-Id antibodies against the monoclonal antibody 3-E7 which recognized opioid peptides (MEO et al. 1983).

One anti-Id sample inhibited the binding of ^{3}H-diprenorphine to the opioid receptor. The cAMP synthesis induced by PGE1 was inhibited by this antibody in a dose-dependent fashion.

Recently, NG and ISOM (1984) obtained anti-Id antibodies against antimorphine antibodies and interactions with opiate receptors were characterized. The anti-Id antibodies cross-reacted with opiate receptors and inhibited the binding of ^{3}H-labeled naloxone to mouse-brain homogenate. The anti-Id serum also produced a biological response: electrically induced contractions were inhibited in a dose-related manner.

3.3.6 *Anti-Idiotypic Antibody Specific for Reovirus Cell Surface*

NOSEWORTHY et al. (1983) have recently characterized a monoclonal anti-Id antibody against an anti-reovirus B cell hybridoma. Thymoma cell lines which bind reovirus 3 were fluorescently stained using the monoclonal anti-Id antibody. The labeling of a subset of murine and rat neuronal cells, in an indirect immunofluorescence assay by the monoclonal anti-idiotypic antibody, confirmed previous studies showing that reovirus binds to the cells. KAUFFMAN et al. (1983) were able to demonstrate an inhibition of the binding of viruses to cell surface receptors by an anti-Id monoclonal antibody. However, the use of anti-Id antibodies to block viral infection failed, probably because just one virus particle is enough to initiate infection.

4 Conclusion

In conclusion, the comparison of the properties of anti-ligand and anti-anti-ligand antibodies with those of receptors for the same ligands and of antibodies against these receptors undoubtedly establishes conceptual if not structural homologies. The ligand-binding site of the two types of macromolecules, antibodies and receptors, share specificity, cross-reactivity, and orders of affinity for series of analogous compounds.

Similarities between membrane-bound receptors and antibodies have been discussed earlier (STROSBERG et al. 1982; STROSBERG 1984), and recent structural evidence concerning the membrane-bound T-cell antigen receptor has indeed confirmed the conservation of the essential features of these similarities: functional domains, variable regions containing the ligand-binding sites, and constant regions that are extracellular, transmembrane and intracytoplasmic.

The immunological analysis of these ligand-binding proteins by the use of antibodies adduces more evidence of their homologies. For example, anti-Id antibodies which inhibit the binding of a hapten to its antibody also block the binding of this same ligand to its physiological, membrane-bound receptor. This inhibition is, in most observed cases, non-competitive, suggesting that the anti-Id although it behaves as analog of the ligand, does not bind at exactly the same site, and/or with the same affinity. For instance, the agonist activity of Ab2B4 for the β-adrenergic receptor on A431 cells is about 25 times lower than that of the synthetic agonist isoproterenol (GUILLET et al. 1985).

Antibodies raised by immunization with affinity-purified receptors may of course be directed against any antigenic part of the molecule. Among those that block ligand binding, quite a few have been described to be competitive inhibitors, but their apparent frequency might be biased by the mode of selection. In many regards the anti-Id antibodies which interact with receptors are quite analogous to the antibodies raised against the receptors and selected for inhibition of ligand binding.

Whether this analogy reflects structural homology between receptor and anti-ligand antibody or molecular mimicry of the anti-Id which adopts, in its active site, a conformation similar to that of the ligand, will remain open for debate as long as structural data are not available.

In this respect it is worth mentioning that the amino acid sequence of the D region of anti-Id antibodies raised against anti-glutamyl-alanyl-tyrosine (GAT) antibodies seem to contain glutamic acid, alanine and tyrosine residues (Fougereau et al. 1985). This would suggest that a quite simple relation might exist between the ligand and the anti-antiligand antibody. Such a correlation will be more difficult to identify when the ligand is not composed of amino acid residues.

Acknowledgments. We wish to thank P. Rossow for critical discussion and C. Peyrat and R. Schwartzmann for help in preparing the manuscript.

References

Atlas D, Steer ML, Levitzki A (1974) Stereospecific binding of propanolol and catecholamines to the β-adrenergic receptor. Proc Natl Acad Sci USA 71:4246–4250

Bona CA, Victor-Kobrin C, Manheimer AJ, Bellon B, Rubinstein LJ (1984) Regulatory arms o the immune network. Immunol Rev 79:26–44

Chamat S, Hoebeke J, Strosberg AD (1984) Monoclonal antibodies specific for β-adrenergic ligands. J Immunol 133:1547–1552

Cherksey BD, Randall BM, Zadunaisky JA (1981) Fluorescence studies of the β-adrenergic receptor topology. Biochemistry 20:4278–4282

Cleveland WC, Wassermann NH, Sarangarayan R, Penn AC, Erlanger BF (1983) Monoclonal antibodies to the acetylcholine receptor by a normally functioning auto-anti-idiotypic mechanism. Nature 305:56–57

Couraud PO, Lu BZ, Schmutz A, Durieu-Trautmann O, Klutchko-Delavier C, Hoebeke J, Strosberg AD (1983a) Immunological studies of β-adrenergic receptors. J Cell Biochem 21:187–193

Couraud PO, Lu BZ, Strosberg AD (1983b) Cyclical anti-idiotypic response to anti-hormone antibodies due to neutralization by autologous anti-anti-idiotypic antibodies which bind hormone. J Exp Med 157:1369–1378

Delavier-Klutchko C, Hoebeke J, Strosberg AD (1984) The human carcinoma cell line A431 possesses large numbers of functional β-adrenergic receptors. FEBS Lett 169:151–155

Durieu-Trautmann O, Delavier-Klutchko C, Hoebeke J, Strosberg AD (1985) Altered binding properties of β-adrenergic receptors and lack of coupling to adenylate cyclase in P815 mastocytoma cells. Eur J Pharmacol 108:133–141

Farid NR, Pepper B, Urbina-Briones R, Islam NR (1982) Biologic activity of anti-thyrotropin anti-idiotypic antibody. J Cell Biochem 19:305–313

Farr RS (1958) A quantitative immunochemical measure of the primary interaction between *I-BSA and antibody. J Infect Dis 103:239–243

Fougereau M, Corbet S, Ollier P, Rocca-Sera J, Roth C, Schoff C, Sommé G, Theze J, Tonnelle C (1985) Structural and genetic basis of the three levels of the idiotypic molecule (Ab1, Ab2 and Ab3) in the GAT systems. Ann Immunol (Paris) 136C:143–156

Guillet JG, Chamat S, Hoebeke J, Strosberg AD (1984) Production and detection of monoclonal anti-idiotypic antibodies directed against a monoclonal anti-β-adrenergic ligand antibody. J Immunol Methods 74:163–171

Guillet JG, Kaveri SV, Durieu O, Delavier C, Hoebeke J, Strosberg AD (1985) β-adrenergic agonist activity of a monoclonal anti-idiotypic antibody. Proc Natl Acad Sci USA 82:1781–1784

Hoebeke J, Vauquelin G, Strosbeg AD (1978) The production and characterization of antibodies against β-adrenergic antagonists. Biochem Pharmacol 27:1527–1532

Hoebeke J, Chamat S, Marullo S, Guillet JG, Strosberg AD (1983) Use of the EIA technique in the functional characterization of anti-drug monoclonal antibodies. In: Avrameas S, Druet P, Masseyeff R, Feldmann G (eds) Immunoenzymatic techniques. Elsevier, Amsterdam, pp 307–311

Homcy CJ, Rockson SG, Haber E (1982) An anti-idiotypic antibody which recognizes the β-adrenergic receptor. J Clin Invest 69:1147–1153

Jerne NK (1974) Towards a network theory of the immune system. Ann Immunol 125C:373–389

Jerne NK (1984) Idiotypic networks and other preconceived ideas. Immunol Rev 79:5–24

Kamel RS, Landon J, Smith DS (1979) Novel ^{125}I-labeled nortriptyline derivatives and their use in liquid-phase or magnetizable solid-phase second-antibody radioimmunoassays. Clin Chem 25:1997–2002

Kauffman RS, Noseworthy JH, Nepom JT, Finberg R, Fields BN, Greene MI (1983) Cell receptors for the mammalian reovirus II. Monoclonal anti-idiotypic antibody blocks viral binding to cells. J Immunol 131:2539–2541

Lefkowitz RJ, Stadel JM, Caron MG (1983) Adenylate cyclase-coupled β-adrenergic receptors. Annu Rev Biochem 52:158–186

Lu BZ, Couraud PO, Schmutz A, Strosberg AD (1984) The internal image of catecholamines: expression and regulation of a functional network. Ann NY Acad Sci 418:240–247

Marasco WA, Becker EL (1982) Anti idiotype as antibody against the formyl peptide chemotaxis receptor of the neutrophil. J Immunol 128:963

Marasco WA, Showell IIJ, Freer RJ, Becker EL (1982) Anti-f Met-Leu-Phe: similarities in fine specificity with the formyl peptide chemotaxis receptor of the neutrophil. J Immunol 128:956

Mason DA, Williams AF (1980) The kinetics of antibody binding to membrane antigens in solution and at the cell surface. Biochem J 187:1–20

Meo T, Gramsch C, Inan R, Hollt V, Weber E, Herz A, Riethmuller G (1983) Monoclonal antibody to the message sequence Tyr-Gly-Gly-Phe of opioid peptides exhibits the specificity requirements of mammalian opioid receptors. Proc Natl Acad Sci USA 80:4084–4088

Morahan G (1983) A sensitive radioimmunoassay for the detection of monoclonal anti-idiotope antibodies. J Immunol Methods 57:165–170

Ng SD, Isom GE (1984) Binding of anti-morphine anti-idiotypic antibodies to opiate receptors. Eur J Pharmacol 102:187–190

Noseworthy JH, Fields BN, Dichter MA, Sobotka C, Pizer E, Perry LL, Nepom JT, Greene MI (1983) Cell receptors for the mammalian reovirus I. Syngeneic monoclonal anti-idiotypic antibody identifies a cell surface receptor for reovirus. J Immunol 131:2533–2538

Schreiber AB, Couraud PO, Andre C, Vray B, Strosberg AD (1980) Anti-alprenolol anti-idiotypic antibodies bind to β-adrenergic receptor and modulate catecholamine-sensitive adenylate cyclase. Proc Natl Acad Sci USA 77:7385–7389

Schulz R, Gramsch C (1984) Anti-idiotypic opioid receptor antibodies. Neuropeptides 5:221–224

Sege K, Peterson PA (1978a) Anti-idiotypic antibodies against anti-vitamin A transporting protein react with pre-albumin. Nature 271:167–168

Sege K, Peterson PA (1978b) Use of anti-idiotypic antibodies as cell-surface receptor probes. Proc Natl Acad Sci USA 75:2443–2447

Shechter Y, Maron R, Elias D, Cohen IR (1982) Autoantibodies to insulin receptor spontaneously develop as anti-idiotypes in mice immunized with insulin. Science 216:542–545

Strosberg AD (1984) Receptors and recognition: from ligand binding to gene structure. Trends Biochem Sci 9:166–169

Strosberg AD, Schreiber AB (1984) Antibodies to receptors and idiotypes as probes for hormone and neurotransmitter receptor structure and function. In: Greaves HF (ed) Receptors and recognition, series B, vol 17. Chapman and Hall, London, pp 15–38

Strosberg AD, Couraud PO, Durieu-Trautmann O, Delavier-Klutchko C (1982) Biochemical and immunological analysis of β-adrenergic receptor adenylate cyclase complexes. Trends Pharmacol Sci 3:282–285

Strosberg AD, Chamat S, Guillet JG, Lavaud B, Emorine L, Hoebeke J (1985) Idiotypy of catecholamine binding proteins. Ann Immunol (Paris) 136C:157–168

Thielemans K, Maloney DG, Meeker T, Fujimoto J, Doss C, Warnke RA, Bindl J, Gralow J, Miller RA, Levy R (1984) Strategies for production of monoclonal anti-idiotype antibodies against human B cell lymphomas. J Immunol 133:495–501

Wasserman NH, Penn AS, Freimuth PI, Treptow N, Wentzel S, Cleveland WL, Erlanger BF (1982) A new route to anti-acetylcholine receptor antibodies and experimental myasthenia gravis. Proc Natl Acad Sci USA 79:4810–4814

Self-Recognition, Auto-Immunity, and Internal Images*

M. ZANETTI and D.H. KATZ

1 Introduction

Although vast knowledge concerning the functional aspects of the immune system has been amassed over the years, contemporary immunology still wrestles with questions concerning the teleologically primary function of the immune system. According to the leading paradigm, the immune system is a constitutive defense system against the microbial world. The demonstrated efficacy of vaccination procedures against virus diseases leaves no doubt that the organism is able to recognize foreign antigens, and thus contributes validity to the paradigm. That the basic function of the immune system could be more fundamental than just constituting a barrier against the microbial world was first suspected by BURNET (1964), who wrote, "It is not difficult to persuade oneself that the development of immunity against pathogenic micro-organisms is of survival value, but for many years I have found this a rather unsatisfying and naive approach. The phenomena of tolerance and of the non-antigenicity of self-components seem to be more basic than those of post-infectious immunity. I cannot conceive that they evolved from an earlier process concerned only with protection against recurrent infection."

Today, there is ample evidence that self-recognition is a continuous and physiological process (KATZ 1978; KATZ and SKIDMORE 1977; KATZ and SKIDMORE 1978; DAUSSET and CONTU 1980). Three large categories of self-moieties are part of this process: polymorphic MHC determinants, nonpolymorphic organ-specific antigens (such as classical auto-antigens), and antigenic determi-

* This is publication no. 67 of the Department of Immunology of the Medical Biology Institute, La Jolla, CA 92037. The work was supported in part by a Grant-In-Aid from QUIDEL. Dr. Maurizio Zanetti is recipient of a Scholar Award from the Leukemia Society of America Inc.

The Department of Immunology, Medical Biology Institute, 11077 North Torrey Pines Road, La Jolla, CA 92037, USA

nants of the variable region (V) of the antibody molecules, e.g., idiotypes (Id). Although much of the available information stems from studies on the Id of antibodies/lymphocytes (for review, see BLASER and DE WECK 1982) which recognize exogenous or synthetic antigens, a prediction made on JERNE'S Id network hypothesis (1974) would be that the immune response to auto-antigens is also regulated by the Id network.

According to Jerne's theory, the immune system is a web of V domains clonally distributed on lymphocytes as antigen receptors (for review, see EICHMANN 1978). The Id system is operational not only when the immune system is challenged by a nominal antigen (immune state), but also during a steady (nonimmune) state. The latter could also be referred to as a state of unresponsiveness or tolerance vis-à-vis self-antigens (JERNE 1974; RAJEWSKY et al. 1983). The question then arises, what are the antigens, other than nominal antigens, that can perturb this network? A large body of experimental data clearly indicates that antibodies to Id determinants (CAZENAVE 1977; WIKLER et al. 1979; RETH et al. 1981) and Id themselves (FORNI et al. 1980) can perturb the network and specifically modify the immune response to nominal antigens.

The available data sufficiently distinguishes between two types of functional networks; suppressive and enhancing. Both could obviously be of relevance for auto-immunity. The first could be primarily involved in the physiological maintenance of unresponsiveness, i.e., tolerance (JERNE 1974; RAJEWSKY et al. 1983). The second could be the underlying critical event through which the immune system loses its ability to down-regulate the continuously emerging auto-immune clones in order to maintain tolerance (ZANETTI et al. 1984b). The latter could be involved in perpetuating auto-immune processes as well. Numerous experiments have shown that anti-Id antibodies can promote the formation of antibodies of predetermined specificity (TRENKNER and RIBLET 1975; BLUESTONE et al. 1981; for review, see SACKS et al. 1983), activate silent clones (HIERNAUX et al. 1981), and induce Id expression even across the barrier of allotype linkage (PENE et al. 1983; MOSER et al. 1983).

It then appears that although in most situations challenge by nominal antigens is the primum movens of an immune response, its control and its fate are actually governed by a series of reactions that are based upon recognition of self, i.e., Id (JERNE 1974; BONA and PERNIS 1984) or Id/MHC (RUBIN and PIERRES 1981; NAGY et al. 1982), by the individual's own immune system. Whether or not one wants to consider self-recognition as the pivotal, regulatory element of the immune system then becomes pleonastic.

Let us consider that the nominal antigen that triggers the immune system is a self-component (auto-antigen). The predictable chain of events is then simplified, since, assuming that Id-anti-Id interactions are the essential feature of regulation of the immune response (JERNE 1974; COUTINHO et al. 1984), the immune system is restricted to self-antiself reactivity. Based on this new paradigm, we have introduced (ZANETTI et al. 1984b) the concept of *auto-immune network* as the comprehensive framework for antiself reactivity based on a Jerne-type Id network. Thus, it is postulated that *responsiveness to auto-antigens is modulated through the positive and negative influences of receptor-antireceptor, self-specific interactions within the idiotypic network.*

To envisage a self-centered immune system, however, one needs to identify those key elements of the immunologic repertoire that can explain both the basic feature of the immune response itself and also, to some extent, the known phenomenology of auto-immune states. The first can be inferred by the fact that auto-antibodies, e.g., antibodies to self-antigens, express rather conserved patterns of idiotypy. This has been already discussed by us (ZANETTI 1983) and others (ROITT et al. 1983) and therefore will not be re-examined here. The second is that not all Id are similarly important to the economy of the immune system. Based on experiments of successful immunization (BONA et al. 1981) of syngeneic mice with purified antibodies of the immunologic cascade (Ab1, Ab2, Ab3) predicted by the network theory, PAUL and BONA (1982) have introduced the concept of regulatory idiotopes. Accordingly, only a limited set of idiotopes within the entire Id repertoire are indeed immunogenic and, more importantly, deemed to play a pivotal role in autologous regulation. The third element, and possibly the most intriguing, is that anti-Id antibodies can resemble antigens (LINDENMANN 1973; JERNE 1974; LINDENMANN 1979; NISONOFF and LAMOYI 1981; URBAIN et al. 1982), mimic antigen functions, and thus induce antibodies of predetermined specificity (for review, see SACKS et al. 1983). Anti-Id antibodies with such properties have been termed homobody (LINDENMANN 1973) or internal image (JERNE 1974). The latter term is the most commonly used.

Several examples of internal image anti-Id antibodies of exogenous antigens have already been described and are certainly discussed elsewhere in this volume. For this reason we will only discuss internal images that refer to antigens of internal origin (self-antigens). Internal images of self-antigens are, clearly, a tautologic concept. An image can only be internal if it is a reflection of an external object. Therefore, we suggest that autologous anti-Id that look like self-epitopes be referred to as mirror images or homobodies. The term internal image should then be used to define standard anti-Id antibodies, i.e., those which mimic exogenous antigens or those which mimic auto-antigens but are made in animals of different species or strain.

In addition to these considerations, one last essential point remains to be made. This pertains to a fundamental characteristic of the auto-immune network – its inherent redundancy. Thus, the phenomenon of auto-immunity is based upon self-recognition events of a twofold nature: (a) recognition and autoreactivity towards auto-antigens, and (b) recognition, reactivity and regulation of self-Id. Obviously, cross-reacting Id, regulatory idiotopes, and mirror-image anti-Id are involved in determining the suppression or enhancement of a perturbed auto-immune network.

2 Perturbation of the Auto-Immune Network Leads to Activation of Autoreactive Clones

Indirect evidence that the Id of auto-antibodies could be directly involved in perturbing the auto-immune network, stems from our own experience during

Table 1. Rabbits immunized with purified auto-antibodies (idiotype) produce auto-antibodies of the same specificity

Immunogen (idiotype)			Serum of immunized rabbits		
Auto-antibody specificity	Origin	Purification procedure	Auto-antibody specificity	Positive (*n*)	Immunized (*n*)
Antithyroglobulin (spontaneous)	Rat serum	Elution from antigen immuno-adsorbent column	Antithyroglobulin	2	2[a]
Antiglomerular basement membrane (spontaneous)	Human kidney	Elution from diseased kidney	Anti-glomerular basement membrane	1	1[b]
Antitubular basement membrane (induced)	Rat kidneys	Elution from diseased kidney	Antitubular basement membrane	3	3[c]
Antithyroglobulin (induced)	Mouse monoclonal antibody	DEAE chromatography	Antithyroglobulin	3	3[d]

Rabbit immunizations were part of individual projects and done with the intent of preparing heterologous anti-idiotypic antibodies. Information concerning immunization schedule and tests used can be found in the pertinent reference

[a] Zanetti and Bigazzi (1981)

[b] Zanetti M, Wilson CB (unpublished results); auto-antibodies to glomerular basement membrane antigens were detected using indirect immunofluorescence

[c] Zanetti et al. (1983b)

[d] Zanetti et al. (1983a)

the preparation of conventional anti-Id antibodies against purified auto-antibodies (M. Zanetti, unpublished results). A surprisingly common feature is that, in every instance, rabbits developed auto-antibodies of the same specificity as the injected auto-antibody. Table 1 indicates the antigen system involved, the origin and the specificity of the injected antibodies, and the type of auto-antibodies detected in the rabbit serum. While recognizing that results from studies using polyclonal serum or tissue-derived auto-antibodies as immunogens are open to criticism, it must be noted that those obtained using monoclonal auto-antibodies – intentionally not purified on antigen immuno-adsorbent columns – strongly suggest activation of the rabbit's own auto-immune network by the injected Id. The possible generation of homobodies and their involvement in initiating auto-immunity is, therefore, only hypothetical.

A similar type of observation, in which internal image antibodies were actually purified and characterized, has been published by Fong et al. (1983). The authors prepared and characterized broadly cross-reactive rabbit anti-Id against a human IgM rheumatoid factor. Anti-Id antibodies recognized, in fact, various polyclonal and monoclonal IgM rheumatoid factors from unrelated individuals. The antigen recognized by the cross-reactive Id was not associated with a particular light- or heavy-chain amino acid sequence, but rather was intrinsic to most proteins with rheumatoid factor activity. F(ab′)2 fragments of rabbit anti-Id thus produced were absorbed onto a rabbit antihuman Fc immuno-adsorbent column from which they could be purified by elution. Collectively, the data suggest that a portion of the produced anti-Id look like the Fc piece of human IgG, i.e., internal image. A subsequent study by the same group (FONG et al. 1984) showed that internal image antibodies could directly activate, in vitro, B lymphocytes from patients with rheumatoid arthritis; albeit no production of rheumatoid factor occurred.

A somewhat analogous experiment was reported by BONA et al. (1982), in which murine anti-Id antibodies were produced against a human IgM monoclonal protein with rheumatoid factor activity. The anti-Id antibodies detected the cross-reactive Id (Wa-IdX) on human rheumatoid factors. These anti-Id were used to prepare murine monoclonal anti(anti-Id) antibodies which were able to bind to human rheumatoid factors and to the Fc fragment of immunoglobulins. Thus, the binding of anti(anti-Id) antibody to the Fc fragment could be explained by the fact that the Id of the human myeloma protein (rheumatoid factor) is mirrored in the Id repertoire of mice, and that monoclonal anti-anti-Id antibodies are themselves topochemical copies of human rheumatoid factor, thereby conferring their ability to bind to the Fc portion of immunoglobulins.

SCHECTER et al. (1982) published an interesting example of perturbation of the auto-immune network whereby immunization with nominal antigen resulted in auto-immune reactivity of pathologic consequences. Mice immunized with insulin not only developed anti-insulin antibodies, but also antibodies that bound to auto-antibodies to insulin and the insulin receptor. The latter were characterized as anti-Id antibodies. Because of their insulin-like activity they are possibly topochemical copies (mirror images) of insulin epitopes. Similar to these results is the demonstration by CLEVELAND et al. (1983) that mice immunized with an agonist of the acetylcholine receptor produce anti-Id antibodies with anti-acetylcholine receptor activity analogous to those found in myasthenia gravis in humans.

Recent experiments in our laboratory (ZANETTI et al. 1984a) have examined whether antibodies to the Id of auto-antibodies can be used as immunogens to perturb the auto-immune network and induce, de novo, auto-antibodies of predetermined specificity thereby substituting for immunization with a classical auto-antigen such as thyroglobulin (Tg). As summarized in Fig. 1, we immunized normal BALB/c mice with purified rabbit antibodies specific for a recurrent idiotype, Id62, borne by a monoclonal antibody to Tg (ZANETTI et al. 1983a). As predicted, the majority (12/13) of anti-Id62-immunized mice, but none of the control mice, produced significant levels of Id62 (Id′). Surprisingly, about half of the anti-Id62-manipulated mice, but none of the control mice, produced auto-antibodies to Tg.

The de novo induction of auto-antibodies of predetermined specificity by anti-Id immuniztion raises a general question as to whether this phenomenon could be due to antibodies carrying the internal image of a nominal auto-antigen. A prerequisite to this possibility would be the demonstration that the same anti-Id antibodies can function as antigen in various species (NISONOFF and LAMOYI 1981). In following this rationale, we used anti-Id62 antibodies to immu-

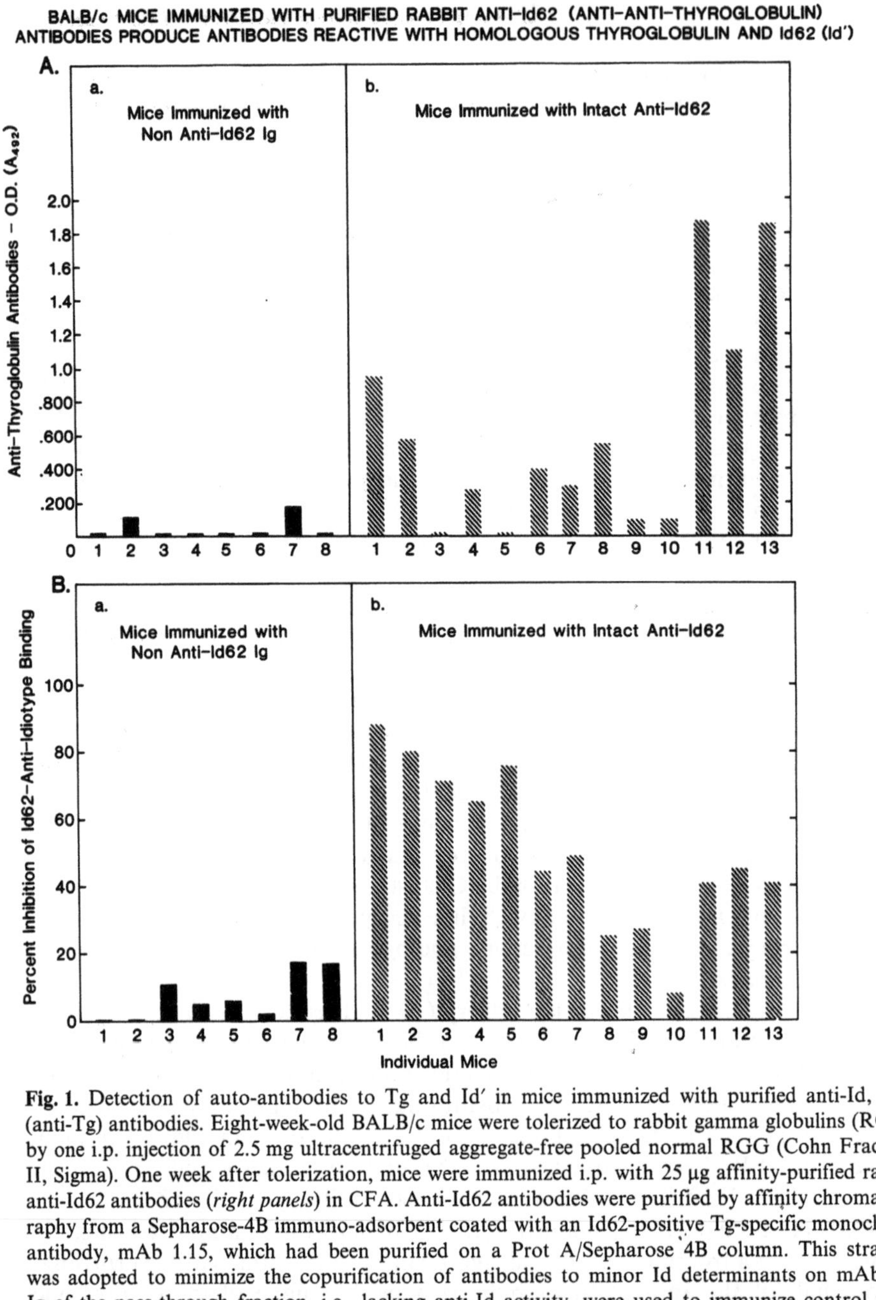

Fig. 1. Detection of auto-antibodies to Tg and Id′ in mice immunized with purified anti-Id, anti (anti-Tg) antibodies. Eight-week-old BALB/c mice were tolerized to rabbit gamma globulins (RGG) by one i.p. injection of 2.5 mg ultracentrifuged aggregate-free pooled normal RGG (Cohn Fraction II, Sigma). One week after tolerization, mice were immunized i.p. with 25 µg affinity-purified rabbit anti-Id62 antibodies (*right panels*) in CFA. Anti-Id62 antibodies were purified by affinity chromatography from a Sepharose-4B immuno-adsorbent coated with an Id62-positive Tg-specific monoclonal antibody, mAb 1.15, which had been purified on a Prot A/Sepharose 4B column. This strategy was adopted to minimize the copurification of antibodies to minor Id determinants on mAb 62. Ig of the pass-through fraction, i.e., lacking anti-Id activity, were used to immunize control mice (*left panels*). Mice were boosted after 2 weeks with an i.p. injection of 5 µg of the same antibody in alum. Sera were individually collected 10 days after the booster injection and tested. Antibodies to Tg (*upper panels*) were detected by ELISA on microtiter plates coated with mouse Tg using a 1:500 serum dilution. The results are expressed as absorbance at 492 nm.

Id′ was detected (*bottom panels*) by competitive inhibition of the ELISA binding of Id62 to microtiter wells coated with 1:4000 dilution of homologous rabbit anti-Id serum by individual mouse sera diluted 1:50. The results are expressed as percent inhibition by comparison with Id62-anti-Id binding in the absence of inhibitor. (Modified from data published in Zanetti et al. 1984a with permission of the publisher)

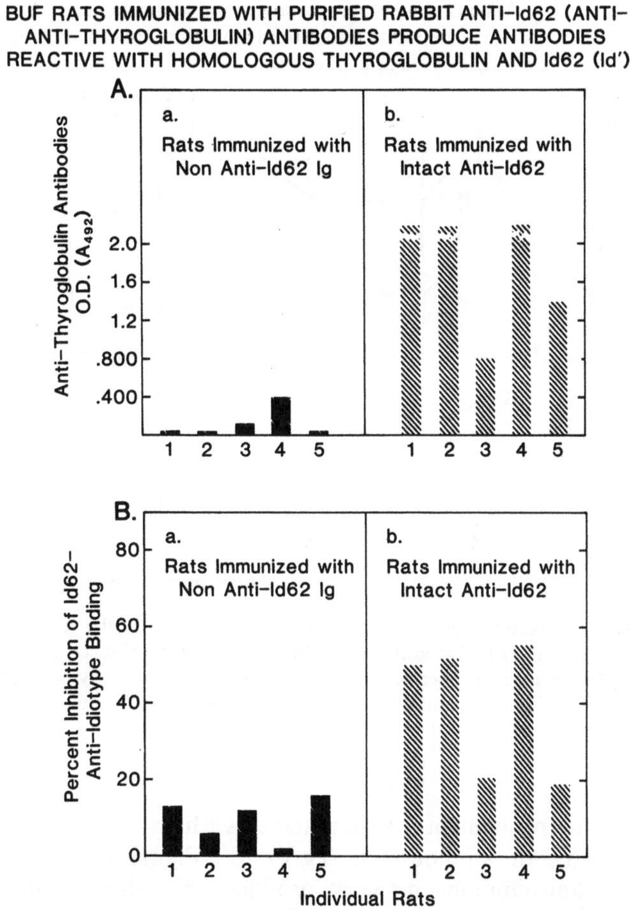

Fig. 2. Detection of auto-antibodies to Tg and Id′ in rats immunized with anti-Id, anti (anti-Tg) antibodies. Two-month-old male BUF rats, which had been previously verified as lacking circulating auto-antibodies to Tg, were tolerized to RGG by one i.p. injection of 10 mg ultracentrifuged aggregate-free pooled normal RGG (Fraction Cohn II, Sigma). One week after tolerization, rats were immunized i.p. with 50 µg anti-Id62 or control Ig in alum. A booster injection of 10 µg of the corresponding antibody was given 2 weeks later. Sera were individually collected 10 days after the booster injection and tested for auto-antibodies to Tg (*upper panels*) by ELISA on microtiter plates coated with rat Tg using a 1:400 serum dilution.

Id′ was detected (*bottom panels*) by competitive inhibition of the ELISA binding of Id62 to microtiter wells coated with 1:4000 dilution of homologous rabbit anti-Id serum by individual rat sera diluted 1:50. The results are expressed as percent inhibition by comparison with Id62-anti-Id binding in the absence of inhibitor. (Modified from data published in Zanetti et al. 1984a with permission of the publisher)

nize species other than mice. Initial studies (Zanetti et al. 1984a) utilized young disease-free Buffalo (BUF) rats, a rat strain that spontaneously develops auto-antibodies to Tg and auto-immune thyroiditis starting from the sixth month of age. Representative results in Fig. 2 show that all five rats immunized with anti-Id62 antibodies produced auto-antibodies to Tg and Id′. In contrast, none

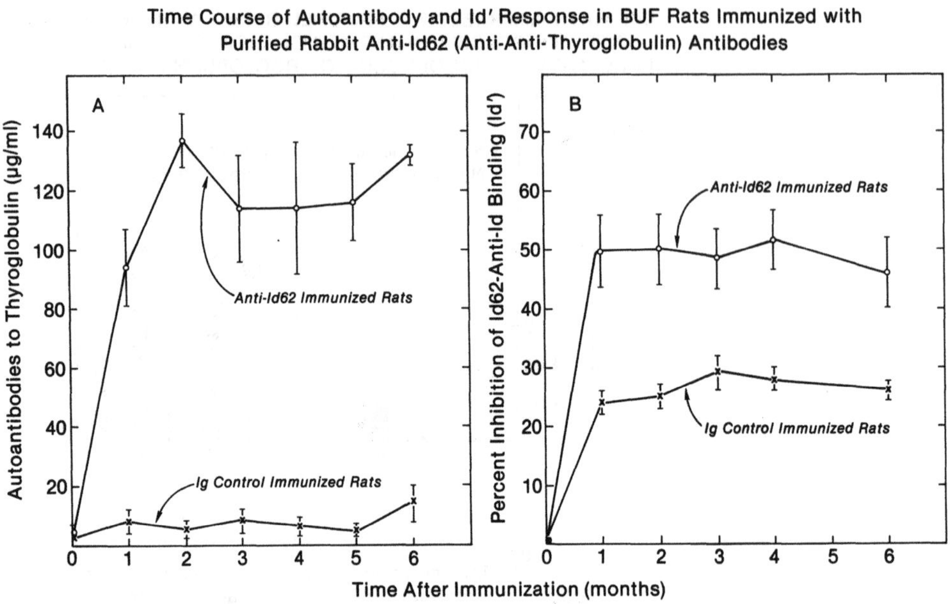

Fig. 3. Persistence of auto-antibodies to Tg and Id′ in the serum of BUF rats immunized with anti-Id, anti (anti-Tg) antibodies. The immunization protocol is as described in Fig. 2. Auto-antibodies to Tg were detected by solid-phase radioimmunoassay performed on microtiter wells coated with rat Tg and quantitated (µg/ml) by plotting the counts per min of experimental sera on a standard curve constructed from known amounts of rat IgG. Id′ was detected as described in Fig. 2

of the five control rats immunized with IgG depleted of anti-Id62 activity produced auto-antibodies to Tg or Id′. Figure 3 shows that the perturbation of the auto-immune network produced by administration of anti-Id antibodies persists for a long time (M. Zanetti, unpublished results). Indeed, both auto-antibodies to Tg and Id′ molecules, remained detectable in significant amounts several months after the initial immunization.

Comparable results were also obtained in the Fischer rat strain which, unlike the BUF strain, is not auto-immune-prone. Immuniztion with anti-Id62 antibodies induced significant levels of auto-antibodies to Tg and Id′ which persisted long after the initial immunization (data not shown). Much to our disappointment, however, we failed to duplicate in guinea pigs the results, as presented above, obtained in mice and rats. In these experiments, two different strains, Hartley and Strain 2 (low and high responders to induction of auto-immune thyroiditis), were used. Nevertheless, as summarized in Table 2, no induction of auto-antibodies to Tg nor Id′ occurred.

The results presented constitute clear evidence that antibodies specific for the Id of auto-antibodies can perturb the natural mechanism of internal homeostasis (tolerance) and induce auto-antibodies of predetermined specificity. Although whether or not the anti-Id antibodies used in our experiments functioned as internal image of a Tg-related epitope is difficult to ascertain, several consider-

Table 2. Immunization with anti-Id62 anti(antithyroglobulin) antibodies elicits auto-antibodies to thyroglobulin and Id′ in various species and strains

Species	Strain	Number of animals which produced auto-antibodies/immunized		Number of animals which produced Id′/immunized	
		Anti-Id62	Ig control	Anti-Id62	Ig control
Mice	BALB/c	7/13	0/13	12/13	0/13
Rat	BUF	5/5	0/5	5/5	0/5
	Fischer	5/5	0/5	5/5	0/5
Guinea Pig	Hartley	0/5[a]	0/5	0/5[b]	0/5
	Strain 2	0/5	0/5	0/5	0/5

Experimental protocols for mice and rats are described in legends to Fig. 1, 2. Guinea pigs were immunized with 50 μg anti-Id antibodies as described for immunization of rats.

[a] Auto-antibodies to Tg were sought by ELISA on microtiter wells coated with guinea-pig Tg

[b] Id′ was detected by competitive inhibition of the ELISA binding of Id62 to microtiter wells coated with 1:4000 dilution of homologous rabbit anti-Id serum by individual guinea-pig sera tested at various dilutions (1:5 to 1:50)

ations suggest that this could be the case. Anti-Id62 antibodies (a) recognize a paratopic (i.e., antigen-combining site associated) Id (Zanetti et al. 1983a), (b) see a recurrent intra/interspecies Id (Zanetti 1983), and (c) promote, as we have documented here, in vivo auto-antibody formation in various species. However, it remains to be determined why immunization of guinea pigs, unlike mice and rats, failed to produce the effect observed in the other species.

To clarify the exact nature of auto-antibodies induced de novo by anti-Id immunization and their relatedness to the antibody at the origin of the immunologic cascade, an Id analysis was made on monoclonal antibodies to Tg prepared from an anti-Id62-manipulated mouse. Auto-antibodies so far examined are idiotypically very similar to the antibody at the origin of the immunologic cascade (Zanetti et al. 1984a). Clearly, this finding could argue against an internal image effect as one would expect internal image antibodies to induce auto-antibodies of predetermined specificity but of various Id. This could be simply explained on the basis of the limited number of clones examined. Alternatively, one could assume that Id62 is a highly conserved Id with regulatory properties in different species. This is not an unlikely hypothesis, as we have recently found that Id62 is expressed on both heavy and light chains of mouse anti-Tg monoclonal antibodies (Zanetti et al. 1985). This could conceivably increase the affinity of anti-Id62 antibodies and render Id62 a preferred Id in immune regulation events. Thus, although an internal image effect would be best evidenced by induction of auto-antibodies with an antigen specificity similar to the original antibody, but with a different Id, the contention of an internal image effect can still be validated if one accepts that regulatory, highly conserved Id are preferentially activated and expressed during perturbation of the auto-immune network.

3 Mirror Images of Self-Antigens on T Cell Receptors

A functional auto-immune network is based upon antibody/cell and cell/cell interactions through complementarity between the Id of the V region of antibodies and cellular receptors. It is generally accepted that autoimmune responses are T cell-dependent phenomena. Similarly, anti-Id responses also require the presence of T cells (KELSOE et al. 1980). Could T cells bearing receptors mimicking self-components (mirror image) play a role in maintaining the immunologic repertoire and regulating the auto-immune network? Let us examine some experiments which may shed light upon these questions. For the sake of convenience, self- and alloantigens will be considered together although, understandably, they are not identical.

The first insight comes from experiments on suppression of alloreactivity in graft vs host (GVH) reactions (BELLGRAU and WILSON 1978; BELLGRAU and WILSON 1979; BELLGRAU et al. 1981; KIMURA and WILSON 1984). As recently conceptualized by WILSON (1984), there are two main features of these experiments. First, there exists in the normal rat a radiosensitive cell whose main function is to prevent the activation of (donor) T cells by host MHC antigenic determinants and consequently the appearance of GVH disease. Second, this "regulatory" T cell can exert its function upon all alloreactive T cells of the same specificity irrespective of the genetic background. An implication of fundamental importance is that T-cell receptors of alloreactive cells bear rather conserved, i.e., monomorphic, "Id" determinants. These results are at variance with previous experiments by BINZ et al. (1977, 1979) who failed to detect such a conserved idiotypy on T-cell receptors with identical allospecificity, but of different genetic origin. Although one could ascribe the discrepancy to the nature of the anti-alloreceptor probe used (anti-"Id" T-cell receptors vs anti-Id antibodies to T-cell receptors), a likely possibility could be that the anti-"Id" T cells of Wilson and colleagues carry the mirror image of a MHC-encoded antigen. Consequently, no polymorphism could be detected on alloreactive receptors.

A formal demonstration that T-cell receptors can look like MHC determinants comes from a recent experiment by SIM et al. (1984), who used a monoclonal anti-I-A antibody to immunize mice from which T cells were selected that proliferated in the presence of the inducing antibody and in the absence of any antigen-presenting cells. Hybrids derived from these T cells produced IL-2 when stimulated by the same anti-I-A antibody, suggesting that this behaves like ligand vis-à-vis structures present on the T-cell hybrid whose activation triggers T-cell function. Furthermore, the original anti-I-A antibody could precipitate, from the putative mirror image-bearing T cell, polypeptide dimers of approximately 40 kD – possibly the T-cell receptor. Finally, from a panel of randomly produced alloreactive anti-I-A T-cell hybrids, the authors were able to isolate clones which interacted directly with the T-cell hybrids bearing receptors expressing the mirror image of a class II antigen.

The importance of these studies lies in the fact that they formally demonstrate that T-T interactions could occur which are uniquely mediated by receptor/antireceptor interactions. A direct consequence of the existence of such a mechanism, if it is indeed physiological, pertains to the maintenance of the T cell repertoire (SIM and AUGUSTIN 1983). Thus, T cells bearing the mirror image of MHC epitopes could provide the stimulus for the selection and conservation of a self-restricted T-cell repertoire at the level of peripheral lymphoid organs (AUGUSTIN et al. 1983).

Collectively, an interesting new concept emerges from the aforementioned experiments, namely that in addition to conventional T cell-bearing receptors with defined self- or allospecificity (T cell of first type), a second type of T cell must be considered which has a regulatory function over the first one. This second T cell type most likely bears a mirror image (imperfect, but functionally active copy) of self MHC antigens as part of its receptor. It is postulated that such a cell will be responsible either for the maintenance of self-tolerance (WILSON 1984) or of the self-restricted T cell repertoire outside the thymic environment (AUGUSTIN et al. 1983).

Experimental evidence for a possible regulatory role of the auto-immune network by T cells with mirror image-carrying receptors, of the type discussed, can already be inferred from known models of auto-immune diseases. For instance, approximately 15% of female BUF rats spontaneously develop auto-immunity to thyroglobulin. However, this incidence of auto-immunity is augmented by neonatal thymectomy or a single exposure to sublethal irradiation in the adult life (M. ZANETTI and P.E. BIGAZZI, unpublished results). In both cases, at least 40% of the rats of both sexes develop auto-antibodies within the 6–8 weeks following the immunologic manipulation. Thus, it appears that the occurrence of auto-immunity to thyroglobulin is under the control of a subset of radiosensitive T lymphocytes, possibly with antiself activity. Similarly, GLAZIER et al. (1983) have shown that irradiated rats repopulated with syngeneic bone marrow cells, and treated daily with cyclosporin A for several weeks, develop a lethal systemic auto-immune disease if cyclosporin A is discontinued. This systemic disease characteristically resembles allogeneic GVH disease. Auto-immunity could be adoptively transferred with T cells from diseased donors into irradiated, but not into non-irradiated, syngeneic recipients. This finding, in conjunction with the fact that cyclosporin A is known to inhibit the development of cytolytic T cells (HESS and TUTSCHKA 1980), suggests that normal rats possess a regulatory T cell with antiself function which is sensitive to cyclosporin treatment and irradiation (WILSON 1984).

4 Selfness and Likeness Are Essential to the Immune System and Establish Its Autonomy

From the foregoing considerations, it can be argued that the concept whereby immunoglobulin V genes have evolved under selective pressures from microbial pathogens is substantially incorrect. Instead, based on the assumption that the immune system's primary activity is self-recognition (KATZ 1978; KATZ and SKIDMORE 1977; KATZ and SKIDMORE 1978; DAUSSET and CONTU 1980; COU-

TINHO et al. 1984; JERNE 1984), it is proposed that the selection and expression of B and T cell repertoires is subjected to a self-recognition process (URBAIN et al. 1981; COUTINHO et al. 1984). While accepting that the complementarity between the two repertoires is guaranteed through a classical "Id" network, one needs to take into account the essential role of internal regulation played by antibody/receptor Id within the concept of a self-centered network, i.e., *selfness*. Since the representation of epitopes of the internal environment (MHC and non-polymorphic organ-specific antigens) on constitutive elements of the immune system, such as antibodies and cell receptors, i.e., *likeness*, is only now being demonstrated (HOLMBERG et al. 1984; SIM et al. 1984), future immunologic theories have to consider selfness and likeness together.

From a merely functional viewpoint one could identify as a distinctive feature of the auto-immune network its physiological, inherent redundancy. Thus, a first order, paratopic, antiself reactivity is followed by a second order, Id antiself reactivity. By virtue of a predictable Id/anti-Id receptor/antireceptor complementarity, any subsequent positive or negative regulation is, a fortiori, based upon self-recognition and self-reactivity. Paradoxically, such a self-centered immune system would be deleterious to the individual unless one accepts that the immune system in its completeness (COUTINHO 1980) also manufactures anti-antiself molecules. This is consistent with a recent speculation by JERNE (1984), who wrote, "I propose that a faithful expression of the anti-self encoding genes does occur to some extent and that evolution, in order to counter stimulation of these undesirable B cells, has favored the emergence of genes which encode complementary elements (e.g., anti-idiotypic antibodies). In ontogeny, self antigens will induce the production of anti-self Ig molecules. But these, binding to self, then invoke the production of anti-idiotypic molecules."

One could further propose that evolution plays another role in the phylogeny of the immune system, namely, the selection of patterns. Let us assume that evolutionary pressures determine (a) epitopic selection on multivalent auto-antigens, i.e., conservation of a limited number of immunogenic epitopes, as well as (b) preferential conservation of certain Id in response to epitopes of the internal environment. Because of the completeness of the internal environment and the redundancy inherent in a self-centered network, certain elements of the auto-immune network are "immortalized" and acquire regulatory properties, i.e., establishment of patterns. Thus, regulatory Id may be viewed as those elements of the auto-immune network preferentially conserved through evolutionary selective pressures.

At present, it is not known whether there exists a structural correlate for these highly conserved forms or, conversely, if they simply reflect a special conformation derived from the rearrangement of V, D, and J gene products. Interestingly, the discovery that Id62, seemingly a regulatory Id, is expressed on both heavy and light chains, suggests that certain, e.g., regulatory, Id may be structurally determined. Conceivably, the concomitant effect of immunodominance by certain Id in concert with the inherent redundancy of a self-centered immune system makes it plausible that immunoglobulin and T-cell receptor Id which look like self-epitopes might be generated. Thus, natural antibodies that resemble self I-A antigens have been recently described (HOLMBERG et al.

1984) and T cells with receptors which display the mirror image of MHC antigens have been constructed (SIM et al. 1984).

Considering that the ability of the immune system to recognize antigens is mirrored at any moment in the immunoglobulin world as negative image (i.e., antibody paratope) and as positive image (i.e., antibody Id) (LINDENMANN 1973), one could then conclude that all Id are mirror images of epitopes (JERNE 1974; COUTINHO 1980). If the latter are self-epitopes, it derives that likeness is part of selfness. Therefore, the immune system does not know whether or not it is dealing with a true, nominal self-antigen or its image. Thus, mirror images of antigens may represent the primitive, imperfect forms that the immune system sees, deals with, and is regulated by. Based on the assumption that the immune system has a propensity to make mirror images of self-antigens, the occurrence of Id as mirror images of self can be regarded not only as conceptually plausible, but also as natural phenomena.

Id which look like self-antigens are, in our opinion, important for the autonomous regulation of the auto-immune network. Let us consider now a hypothetical scenario, based upon a functional connectivity created by mirror images, which may pertain to the pathophysiology of auto-immunity. Auto-immune diseases are often chronic processes whose mechanisms are still poorly understood. We have recently proposed (ZANETTI et al. 1984b) that autoreactive clones can be activated not only by modified self or cross-reacting bacterial antigens, but also by anti-Id specific for receptors on autoreactive clones, though their generation may be the fortuitous product of activation through a parallel set (JERNE 1974). Here, we suggest that mirror images of auto-antigens may play a pivotal role in the maintenance of an auto-immune state once self-tolerance has been broken by other means. T-helper cell-dependent phenomena of B-cell activation, e.g., the auto-immune state, require interactions between receptors on T-helper cells and on B lymphocytes which, conventionally, are linked through nominal antigens (MITCHISON 1971). A threshold, constitutive existence of mirror images of self-epitopes within the internal environment and their possible increment following perturbation of the auto-immune network by whatever means, would provide an alternative, autonomous way through which a functional intercellular connectivity can be realized. Thus, Id on immunoglobulins or T-cell receptors which look like epitopes of nominal auto-antigens could easily function as the silent link between the immunologic compartments and cellular elements. Furthermore, antibody-homobody complexes (LINDENMANN 1973) or bifunctional antibodies (RUBINSTEIN et al. 1983), e.g., antibodies which with their paratope stimulate clones bearing complementary Id (classical anti-Id) and with their Id (i.e., mirror image) activate clones with receptors for self-epitopes, could also be hypothesized. Collectively, these mechanisms would be sufficient to ensure direct productive interactions between B and T lymphocytes.

Clearly, future research will have to determine experimentally which, if any, of these assumptions are correct. For the time being, mirror images of self-antigens serving as a functional bridge between T and B lymphocytes may be regarded as candidates for establishing the autonomy of the immune system and possibly explaining the persistence of many auto-immune responses associated with diseases in humans.

Acknowledgments. The authors wish to thank the excellent assistance of J. Czarnecki and B. Burgess in preparing this manuscript.

References

Augustin AA, Sim GK, Bona CA (1983) Internal images of antigens within the immune network. Surv Immunol Res 2:78–87

Bellgrau D, Wilson DB (1978) Immunological studies of T-cell receptors. I. Specifically induced resistance to graft-versus-host disease in rats mediated by host T-cell immunity to alloreactive parental T cells. J Exp Med 148:103–114

Bellgrau D, Wilson DB (1979) Immunological studies of T-cell receptors. II. Limited polymorphism of idiotype determinants on T-cell receptors specific for MHC alloantigens. J Exp Med 149:234–243

Bellgrau D, Smilek D, Wilson DB (1981) Induced tolerance in F_1 rats to anti-major histocompatibility complex receptors on parental T cells. Implication for self tolerance. J Exp Med 153:1660–1665

Binz H, Wigzell H (1977) Antigen-binding idiotypic T lymphocyte receptors. In: Stutman O (ed) Contemporary Topics in Immunology, vol 7. Plenum, New York, pp 113–117

Binz H, Frishknecht H, Shen FW, Wigzell H (1979) Idiotypic determinants on T-cell subpopulations. J Exp Med 149:910–922

Blaser K, De Weck AL (1982) Regulation of the IgE antibody response by idiotype-anti-idiotype network. Prog Allergy 32:203–264

Bluestone JA, Sharrow SO, Epstein SL, Ozato K, Sachs DH (1981) Induction of anti-H-2 antibodies without alloantigen exposure by *in vivo* administration of anti-idiotype. Nature 291:233–235

Bona CA, Pernis B (1984) Idiotypic Networks. In: Paul WE (ed) Fundamental Immunology. Raven, New York, pp 577–592

Bona CA, Herber-Katz E, Paul WE (1981) Idiotype-anti-idiotype regulation. I. Immunization with a levan-binding myeloma protein leads to the appearance of auto-anti-(anti-idiotype) antibodies and to the activation of silent clones. J Exp Med 153:951–967

Bona CA, Finley S, Waters S, Kunkel HG (1982) Anti-immunoglobulin antibodies. III. Properties of sequential anti-idiotypic antibodies to heterologous anti-γ globulins. Detection of reactivity of anti-idiotype antibodies with epitopes of Fc fragments (homobodies) and with epitopes and idiotopes (epibodies). J Exp Med 156:986–999

Burnet FM (1964) Immunological recognition of self. Nobel lecture (december 12, 1960) In: Nobel Lectures in Physiology or Medicine 1942–1962. Elsevier, Amsterdam, pp 689–701

Cazenave PA (1977) Idiotypic-anti-idiotypic regulation of antibody synthesis in rabbits. Proc Natl Acad Sci USA 74:5122–5125

Cleveland WL, Wassermann NH, Sarangarajan R, Penn AS, Erlanger BF (1983) Monoclonal antibodies to the acetylcholine receptor by a normally functioning auto-anti-idiotypic mechanism. Nature 305:56–57

Coutinho A (1980) The self-nonself discrimination and the nature and acquisition of the antibody repertoire. Ann Immunol (Paris) 131D:235–253

Coutinho A, Forni L, Holmberg D, Ivars F, Vaz N (1984) From an antigen-centered, clonal perspective of immune responses to an organism-centered, network perspective of autonomous activity in a self-referential immune system. Immunol Rev 79:151–168

Dausset J, Contu L (1980) Is the MHC a genral self-recognition system playing a major unifying role in an organism? Hum Immunol 1:15–17

Eichmann K (1978) Expression and function of idiotypes on lymphocytes. Adv Immunol 26:195–254

Fong S, Gilbertson TA, Carson DA (1983) The internal image of IgG in cross-reactive anti-idiotypic antibodies against human rheumatoid factors. J Immunol 131:719–723

Fong S, Gilbertson TA, Chen PP, Vaughan JH, Carson DA (1984) Modulation of human rheumatoid factor-specific lymphocyte response with a cross-reactive anti-idiotype bearing the internal image of antigen. J Immunol 132:1183–1189

Forni L, Coutinho A, Kohler G, Jerne NK (1980) IgM antibodies induce the production of antibodies of the same specificity. Poc Natl Acad Sci USA 77:1125–1128

Glazier A, Tutschka PJ, Farmer ER, Santos GW (1983) Graft-versus-host disease in cyclosporin A-treated rats after syngeneic and autologous bone marrow reconstitution. J Exp Med 158:1–8
Hess AD, Tutschka PJ (1980) Effect of cyclosporin A on human lymphocyte responses *in vitro*. I. CsA allows for the expression of alloantigen activated suppressor cells while preferentially inhibiting the induction of cytolytic effector lymphocytes in MLR. J Immunol 124:2601–2608
Hiernaux J, Bona C, Baker PJ (1981) Neonatal treatment with low doses of anti-idiotypic antibody leads to the expression of a silent clone. J Exp Med 153:1004–1008
Holmberg D, Forsgreen S, Forni L, Ivars F, Coutinho A (1984) Idiotypic determinants of natural IgM antibodies that resemble self Ia antigens. Proc Natl Acad Sci USA 81:3175–3179
Jerne NK (1974) Towards a network theory of the immune system. Ann Immunol (Paris) 125:373–389
Jerne NK (1984) Idiotypic network and other preconceived ideas. Immunol Rev 79:5–24
Katz DH (1978) Self-recognition as a means of cell communication in the immune system. In: Curtis ASG (ed) Cell-cell recognition. Cambridge University Pess, New York, pp 411–427
Katz DH, Skidmore BJ (1977) Self recognition as the predominant mechanism for communication in the immune system. In: Progress in Immunology, vol 3, TE Mandel, C Cheers, CS Hosking, IFC Mc Kenzie, GJV Nossal, eds. Aust Acad Sci, Canberra, pp 322–328
Katz DH, Skidmore BJ (1978) Self-recognition as a means of cell communication in biological systems: the immune system is not an exception to the general rule: In: Lerner RA, Bergsma D (eds) The Molecular Basis of Cell-Cell Interaction, vol 14. Liss, New York, pp 327–341
Kelsoe G, Isaak D, Cerny J (1980) Thymic requirement for cyclical idiotypic and reciprocal anti-idiotypic immune responses to a T-independent antigen. J Exp Med 151:289–300
Kimura H, Wilson DB (1984) Anti-idiotypic cytotoxic T cells in rats with graft-versus-host disease. Nature 308:463–464
Lindenmann J (1973) Speculations on idiotypes and homobodies. Ann Immunol (Paris) 124:171–184
Lindenmann J (1979) Homobodies: do they exist? Ann Immunol (Paris) 130:311–318
Mitchison NA (1971) The carrier effect in the secondary response to hapten-protein conjugates. II. Cellular cooperation. Eur J Immunol 1:18–27
Moser M, Leo O, Hiernaux J, Urbain J (1983) Idiotypic manipulation in mice: BALB/c mice can express the cross-reactive idiotype of A/J mice. Proc Natl Acad Sci USA 80:4474–4478
Nagy ZA, Elliott BE, Carlow DA, Rubin B (1982) T cell idiotypes recognizing self-major histocompatibility complex molecules: H-2 specificity, allotype linkage, and expression on functional T cell populations. Eur J Immunol 12:393–400
Nisonoff A, Lamoyi E (1981) Implications of the presence of an internal image of the antigen in anti-idiotypic antibodies: possible application to vaccine production. Clin Immunol Immunopathol 21:397–406
Paul WE, Bona C (1982) Regulatory idiotopes and immune networks: a hypothesis. Immunol Today 3:230–234
Pene J, Bekkhoucha F, Desaymard C, Zaghouani H, Stanislawski M (1983) Induction of a cross-reactive idiotype dextran-positive antibody response in two IgH-C^b mouse strains treated with anti-J558 cross-reactive idiotype antibodies. J Exp Med 157:1573–1593
Rajewsky K, Takemori T, Muller LE (1983) Self-tolerance through idiotype suppression. In: Yamamura Y, Tada T (eds) Progress in Immunology V. Academic Press, Orlando FL, pp 533–542
Reth M, Kelsoe G, Rajewsky K (1981) Idiotypic regulation by isologous monoclonal anti-idiotope antibodies. Nature 290:257–259
Roitt IM, Male DK, Cooke A, Lydyard PM (1983) Idiotypes and autoimmunity. Springer Semin Immunopathol 6:51–66
Rubin B, Pierres M (1981) Anti-MHC antibody idiotypes as important elements in immunoregulation. Immunol Today 2:208–209
Rubinstein LJ, Goldberg B, Hiernaux J, Stein KE, Bona CA (1983) Idiotype-anti-idiotype regulation. V. The requirement for immunization with antigen or monoclonal anti-idiotypic antibodies for the activation of $\beta 2 \rightarrow 6$ and $\beta 2 \rightarrow 1$ polyfructosan reactive clones in BALB/c mice treated at birth with minute amounts of anti-A48 idiotype antibodies. J Exp Med 158:1129–1144
Sacks DK, Kelsoe GH, Sachs DH (1983) Induction of immune responses with anti-idiotypic antibodies: implication for the induction of protective immunity. Springer Semin Immunopathol 6:79–97
Shechter Y, Maron R, Elias D, Cohen IR (1982) Autoantibodies to insulin receptor spontaneously develop as anti-idiotypes in mice immunized with insulin. Science 216:542–545

Sim GK, Augustin AA (1983) Internal images of major histocompatibility complex antigens on T-cell receptors and their role in the generation of the T-helper cell repertoire. Ann NY Acad Sci 418:272–281

Sim GK, Nelson JE, Augustin AA (1984) Some T cell receptor idiotypes are internal images of MHC epitopes. In: Sercarz E, Cantor H, Chess L (eds) Regulation of the immune system, vol 18. Liss, New York

Trenkner E, Riblet R (1975) Induction of antiphosphorylcholine antibody formation by anti-idiotypic antibodies. J Exp Med 142:1121–1132

Urbain J, Wuilmart C, Cazenave P-A (1981) Idiotypic regulation in immune networks. In: Inman FP, Mandy WJ (eds) Contemp Topics Molec Immunol, vol 8. Plenum, New York, pp 113–148

Urbain J, Slaoui M, Leo O (1982) Idiotypes, recurrent idiotypes and internal images. Ann Immunol (Paris) 133D:179–189

Wilson DB (1984) Idiotypic regulation of self-rective and allo-reactive T cells in autoimmunity and graft-versus-host disease. Immunol Today 5:228–230

Wikler M, Franssen JD, Collignon C, Leo O, Mariame B, Van de Walle P, De Groote D, Urbain J (1979) Idiotypic regulation of the immune system. Common idiotypic specificities between idiotypes and antibodies raised against anti-idiotypic in rabbits. J Exp Med 150:184–195

Zanetti M (1983) Anti-idiotypic antibodies and autoantibodies. Ann NY Acad Sci 418:363–387

Zanetti M, Bigazzi PE (1981) Anti-idiotypic immunity and autoimmunity. I. *In vitro* and *in vivo* effects of anti-idiotypic antibodies to spontaneously occurring autoantibodies to rat thyroglobulin. Eur J Immunol 11:187–195

Zanetti M, De Baets M, Rogers J (1983a) High degree of idiotypic cross-reactivity among murine monoclonal antibodies to thyroglobulin. J Immunol 131:2452–2457

Zanetti M, Mampaso F, Wilson CB (1983b) Anti-idiotype as a probe in the analysis of autoimmune tubulointestitial nephritis in the Brown Norway rat. J Immunol 131:1268–1273

Zanetti M, Rogers J, Katz DH (1984a) Induction of autoantibodies to thyroglobulin by anti-idiotypic antibodies. J Immunol 133:240–243

Zanetti M, Rogers J, Katz DH (1984b) Induction of autoreactivity by anti-idiotypic antibodies. In: Sercarz E, Canter H, Chess L (eds) Regulation of the immune system, vol 18. Liss, New York, pp 893–907

Zanetti M, Liu F-T, Rogers J, Katz DH (1985) Heavy and light chains of a mouse monoclonal autoantibody express the same idiotype. J Immunol, in press

The Idiotypic Network: Order from the Beginning or Order out of Chaos?

J. URBAIN, M. BRAIT, C. BRUYNS, C. DEMEUR, P. DUBOIS, M. FRANCOTTE, J.-D. FRANSSEN, J. HIERNAUX, O. LEO, J. MARVEL, P. MEYERS, M. MOSER, M. SLAOUI, J. TASSIGNON, G. URBAIN-VANSANTEN, A. VAN-ACKER, M. WIKLER, F. WILLEMS, and C. WUILMART

1 Remembrances

Some years ago, we came upon the concept of the idiotypic network (JERNE 1974) on the basis of some unexpected findings and because of a prejudice in favor of germ-line theories (URBAIN et al. 1972; URBAIN 1974; URBAIN 1976). At that time, there was controversy between adherents of the somatic mutation theory and "hard-nosed" germ liners.

The crux of germ-line theory is selection pressure. How does a species maintain genes, most of which are not likely to be used in the lifetime of one individual (antibodies against crocodila albumin, new synthetic antigens, etc.)? How could we explain the most astonishing property of the immune system, the idiotype (Id) phenomenon?

When considering the Id phenomenon, one is immediately confronted with the problem of available and potential repertoires.

Our work has been deeply influenced by a paper entitled, "What precedes clonal selection?", where N. JERNE (1972) wrote:

"If, conceptually, we were to place the potential repertoire in the germ line – that is, if we assume that all structural V genes for the antibodies that an individual may potentially express are already present in the DNA of the zygote – then we would be tempted to conclude that only a small fraction of these are actually expressed in the available repertoire of an individual. Otherwise,

Université Libre de Bruxelles, Laboratoire de Physiologie Animale, Département de Biologie Moléculaire, 67, rue des Chevaux, B-1640 Rhode-St-Genèse

Current Topics in Microbiology and Immunology, Vol. 119

genetically identical or related animals would be expected to produce, at least in part, identical antibody molecules to the same antigen.

This repertoire restriction appears to make a germ-line hypothesis untenable. We must admit that the number of V genes required to encode an available antibody repertoire is already uncomfortably large, if it must be located in the germ-line genome. The situation becomes worse if we consider that a germ-line theory would require the presence, in the genome of the zygote, of the entire potential repertoire, which is far larger. A collection of genes can be kept intact in evolution only if each gene is used and if its absence impairs survival to some degree. It is hard to believe that the presence of every gene in the large set required to encode the potential repertoire is essential. We cannot be quite certain of this, however, for even if a given light-chain V gene is not expressed in combination with any of several heavy-chain V genes, it may find expression in combination with other heavy-chain V genes. In spite of this consideration, the fact that the potential repertoire is much larger than the available repertoire suggests that many V genes can be dispensed with. As unessential genes will be lost by mutation, it appears that we must look for a different genetic solution to the problem of the repertoire. Such a solution must start with a number of essential V genes for antibodies in the zygote. This number must be much smaller than the number of such genes expressed by the lymphocyte populations present in a set of genetically identical immunocompetent animals."

Upon reading Jerne's paper, we wondered whether it would be possible to imagine a germ line theory, despite the absence of a sufficient external selective pressure. We thought that the insufficient external selective pressure could be replaced by an internal selective pressure as diverse as the world of antigens. Therefore, the germ line could encode both Id and anti-Id (germ-line circles, or order from the beginning). These Id and anti-Id can be conserved because they are used in clonal interactions. This hypothesis implies that the potential Id repertoire should be more or less similar in all individuals from a given species or even between different species. If this is the case, it should be possible to induce, at will, "silent" Id.

In this hypothesis, Abl corresponds to antiself or anti-nonself antibodies (as does Ab2).

In a very recent paper, Jerne (1984) has proposed a more dramatic version of germ-line circles, by suggesting that the germ line is mainly encoding for antiself antibodies and for anti-Id antibodies to these antiself antibodies. This implies that somatic mutations play a major role in the establishment of the antinonself repertoires.

2 The Idiotypic Cascade – or Antibodies Against Antibodies Against Antibodies...

Jerne has proposed that the number of antibodies produced against any given antigen is regulated by antibodies against those antibodies. The second order antibodies (Ab2) recognize Id specificities of the Id (Ab1). These markers are near or within the binding site; however, they are not necessarily the same structures as those involved in antigen recognition.

CAZENAVE (1977), at the Institut Pasteur, and our group at the Université de Bruxelles (URBAIN et al. 1977) became interested in whether the production of certain antibodies could be artificially regulated by interfering with the natural network. In both laboratories, a rabbit (rabbit I) was immunized with a specific antigen (ribonuclease in Cazenave's experiment and *Micrococcus lysodeikticus* in ours). The antibodies produced, designated Ab1, were then used to immunize other rabbits (rabbits II), which produced antibodies against the Id on the Ab1. These anti-Id antibodies were designated Ab2. The Ab2 antibodies were then injected into still other rabbits (rabbits III), which produced antibodies, Ab3, against Ab2 Id. Finally, rabbits III were immunized with the original antigen ribonuclease or micrococcus.

BOTH we and CAZENAVE theorized that rabbit I, confronted with the original antigen, would be capable of responding with the production of a wide range of different antibodies, with an equally wide range of different Id. The particular antibody actually produced in large quantity would depend on which one escaped suppression by the regulatory network, consisting of antibodies against the Id of other antibodies, and containing some antibodies against the Id of Ab1, but not enough to suppress its production. Rabbits II function simply as a source of larger numbers of antibodies to Ab1. Rabbits III are crucial, since they produce antibodies against the antibodies that suppress Ab1 production.

Suppression of suppressors of an antibody would then be expected to result in more of that antibody. Thus, large quantities of antibodies bearing the Ab1 Id were expected from the immunization of rabbits III with ribonuclease or micrococcus.

Under normal circumstances, one would not expect to find antibodies of the same Id predominating in the immune response of two different rabbits. We tested 60 non-immunized rabbits with micrococcus and only one of them produced antibodies with the Id of Ab1. However, all rabbits III produced antibodies with the Ab1 Id. Thus, we seem to have poked a hole in the anti-Id network, allowing enhanced production of the Ab1 Id, which would otherwise have been kept to very low levels or not produced at all. It has now become clear that immunization with Ab2 can induce four subsets of Ab3 (SCHNURR 1981). The first is comprised of Id^-Ag^- antibodies which recognize Ab2 exclusively, share no Id (the determinants formed by the variable regions of the heavy and/or light chains of an antibody molecule), and are unable to bind to antigen. These antibodies are serologically and genetically unrelated to Ab1, although this does not exclude a physiological role for them (see POLLOCK and KEARNEY 1984). The second subset shares Id with Ab1 but does not bind antigen (Id^+Ag^-). In most instances, this subset is largely represented in the total Ab3 population. The existence of this subset was first demonstrated by the fact that Ab4 antibodies recognize Ab1 antibodies. The third subset is very similar to Ab1 (Id^+Ag^+) because it shares Id and binds antigen. If the Ab2 used for immunization contains Ab2β antibodies, i.e., internal images, the fourth Ab3 subset appears, which is predominantly Id^-Ag^+ (JERNE et al. 1982).

The induction of "silent" Id can be achieved in more physiological conditions by maternal immunoglobulins. In these experiments, female rabbits are injected with a randomly chosen Ab2 to raise Ab3 antibodies and are mated with nonimmunized males. Both the offspring at 2 months of age and the mother are then injected with the original antigen (the one used to induce Id Ab1). The "silent" Id (Ab1-like) appears in the mother and in 50% of the offspring. The reverse experiment, i.e., males producing Ab3 antibodies crossed with nonimmunized females, results in the expression of "silent" Id only in the father and not in the offspring. Simultaneous immunization of a rabbit with six different Ab2 directed against six Ab1 (noncrossreactive Id but specific for the same antigen) enables the rabbit to produce simultaneously six noncrossreactive Id. Thus, the available repertoire can be tremendously expanded at will (WIKLER and URBAIN 1984).

However, induction of a new silent Id after the rabbits have been injected with antigen or manipulated by a previous Id cascade is impossible; once the "selection" is completed, either by the immune system itself or as predetermined by experimental manipulation, a stable steady state is reached. We call this phenomenon "original idiotypic sin." (WIKLER and URBAIN, manuscript in preparation).

3 The Arsonate Saga

To go further, we needed a model system for which detailed genetic and structural information was available. At the time we began these studies, it was known that in the A/J strain of mice, the recurrent Id CRI_A appearing after immunization with arsonate coupled to KLH (Ars-KKH), is a large family of related but not identical immunoglobulins (KUETTNER et al. 1972; CAPRA et al. 1982; GREENE et al. 1982). In fact, the Id, like most major Id, e.g., Np, T15, results from the transcription and translation of a rearranged gene composed of germ-line segments, namely V_H id crll-DF1 16.1 and JH_2 (SIEKEWITZ et al. 1982). Thus, the recurrent Id is largely monoclonal and the diversity stems from somatic mutations, the latter accounting for the "microheterogeneity" of the Id. In most recurrent Id, expression is linked to the Igh locus. Normally, this Id, is absent in strains that do not possess the Igh^d or Igh^e haplotype. Balb/c is a typical nonexpressor strain. Furthermore, from genetic studies, Gefter and his group have shown that the V segment Id crll from A/J mice seems to be absent in the genome of Balb/c mice. Thus, at first glance, the situation seems to be simple, i.e., recurrent Id are expressed when the suitable germ line is present, and remain "silent" in other strains, where the germ-line corresponding segments are absent. DNA polymorphism and clonal selection theory would then explain everything, the network theory becoming mere intellectual fantasy. However, our attempts to induce CRI_A-like molecules in Balb/c mice by the Id cascade clearly resulted in the induction of CRI_A-like antiarsonate antibodies in every Balb/c mouse after immunization with rabbit Ab2, followed by antigen (MOSER et al. 1983). Such results were obtained previously by LUCAS and HENRY, who simply injected arsonate coupled to *Brucella abortus* (LUCAS and HENRY 1982). Since all these initial experiments were done using polyclonal antisera, we decided to conduct a more detailed analysis using a panel of Ab2 monoclonal antibodies, directed against the germ-line recurrent Id Ab1 3665 (kindly given by M. GEFTER) of the A/J mouse strain. These monoclonal Ab2 were used to compare the germ-line encoded recurrent Id of A/J mice (CRI_A) and the "induced" CRI_A-like Id obtained in each manipulated Balb/c mouse. The results were clear: four out of the five idiotopes recognized by our monoclonal Ab2, which recognize distinct Id on 3665, were shared by 3665 and by the induced Id of manipulated Balb/c mice (LEO et al. 1985). Using this panel of Ab2, we analyzed the sera of Balb/c mice immunized only with antigen to determine whether these mice did not, in fact, ever express the Id of A/J mice.

Surprisingly, 20% of Balb/c mice immunized with Ars-KLH expressed very low levels of CRI_A-like anti-arsonate antibodies. Furthermore, 2%–3% of the Balb/c mice immunized with antigen only were able to synthesize appreciable amounts of antibodies (100 μg) expressing the major crossreactive Id of the A/J strain. The CRI_A-like Id, which appears in few Balb/c mice after antigen immunization alone behave like an Id "*à la* Oudin." In these idiotypically crossreactive Id, one idiotope, which has been mapped near the second hypervariable region of the Id crll segment of the A/J strain, was consistently missing. Moreover, the CRI_A-like antibodies from immunized or manipulated Balb/c mice were nonreactive with anti-91A3 antiserum, which recognizes specifically the V subgroup of the 3665 (the germ-line product of the A/J strain).

Two major points emerge from these results. First, the results are essentially rediscoveries of findings by KUNKEL et al. (1963) and OUDIN and MICHEL (1963). After all, Id were detected first in humans and rabbits. KUNKEL et al. showed that any reaction between an anti-Id antiserum and individual antigenic specificities of a myeloma protein can be inhibited by normal human immunoglobulins, provided that the inhibitor is present at sufficient concentration. OUDIN and BORDENAVE (1971) showed that the rules of idiotypy are not absolute and that most Id systems are characterized by a frequency of crossreactions ranging from 2%–5% (the estimate depends on the sensitivity of the test, as elegantly shown by T. KINDT et al. 1973). Second, the results on the expression of CRI_A-like molecules in Balb/c mice provide a system that allows comparisons, at the molecular level, of a recurrent Id (the CRI_A of A/J), an Id that appears after immunization with antigen alone (the CRI_A-like Id appearing in Balb/c mice) and an Id induced by the Id cascade. Therefore, we prepared CRI_A-like monoclonal Ab1 antibodies from Balb/c mice immunized with antigen alone and from "manipulated" Balb/c mice. These monoclonal antibodies have been sequenced by D. Capra's group (MEEK et al. 1984). For the sake of simplicity, we refer to the three groups of strongly crossreactive Id as: germ-line encoded (CRI of A/J mice, 3665; "spontaneous" (22B5 originating from Balb/c mice immunized with antigen only); and "induced" (9A5 originating after the Id cascade). The spontaneous and induced Id are nearly identical (despite their derivation from two different genes). As far as we know, the Id cascade does not select mutants but allows the expression of a silent part of the potential repertoire (discussed in detail in LEO et al. 1985).

The amino acid sequence of the germ-line encoded major CRI from A/J mice differs vastly from its counterpart in Balb/c (22B5 and 9A5), with only 50% homology. In particular, the V_H segment of crossreactive antibodies from Balb/c mice are completely different and belong to a *new* subgroup or a *new* family. Therefore, the counting of genes, as proposed by BRODEUR and RIBLET, is really an underestimate of the germ-line potential.

The D segments of all three Id are remarkably similar, as are their light chains. Interestingly, there are groups of molecules that are strongly idiotypically crossreactive, while differing drastically in amino acid sequences. T-cell receptors and immunoglobulins also differ (50%), but they can share Id specificities. Also, it is interesting that serological and structural information are reconciled. These results are in agreement with the DNA data on Gefter's group and, furthermore, were predicted using detailed serological analysis.

From the above discussion, it would appear that the D segment is a key element for the CRI_A Id. But caution must be exercised in Id mapping (CAPRA and FOUGEREAU 1984). After all, an Id is first defined operationally, i.e., it exists because it is recognized by an anti-Id serum. Polyclonal anti-Id sera vary in different bleedings from the same animal and also vary among rabbits. Furthermore, an Id is a family of immunoglobulins sharing some but not all Id specificities (idiotypes). Any Id determinant is a mosaic of idiotopes. We can say only that some segments of the molecule are necessary for the expression of one idiotope, although even the modification of one key element, leading to the loss of one idiotope, can be "complemented" by some other changes in other parts of the same molecule. We also know that X-ray crystallography is not the final solution in defining the CRI. However, two points should be stressed: (a) It appears that the D segments, in conjunction with "permissive" other V_H segments and V_L segments, are part of the CRI_A; and (b) recent molecular studies of T. Manser using a very elegant method (NEAR et al. 1985, lysis hybridization and dot-blot analysis), strongly suggest that "one particular combination of gene segments" – V_H Id crll-DF1 16.1-JH_2 – is "singled out" in A/J mice during the immune response of A/J mice (MANSER et al. 1985). Our studies show that the greatest similarity between the crossreactive A/J and Balb/c antibodies occurs in the D segment, but that the D segment in Balb/c mice is associated with a new family of V segments.

Why is the CRI_A dominant in the immune response of A/J mice immunized with Ars-KLH? Is it simply a matter of antigen affinity, the capacity to generate "useful" somatic variants? Why does the CRI_A-like Id of Balb/c behave as an Id "*à la* Oudin," except in manipulated Balb/c mice? Clearly, the dominance of the recurrent CRI_A Id in A/J mice is not due to gene amplification. One V gene, or at most a few, are involved. As mentioned, a single combination of gene segments is "singled out" of a large population of gene segment combinations that are equally represented in the preimmune repertoire.

The dominance cannot be explained by the presence of a large proportion of relevant precursors in the preimmune repertoire, a conclusion based on results from three sets of experiments, using completely different methods: the splenic focus assay (SIGAL 1982), limiting dilution analysis after polyclonal activation (SLAOUI et al. 1984), and a direct molecular approach (WYSOCKI et al. 1984).

Table 1. Antibody and idiotypic response of A/J mice immunized with arsonate[a]

Donor cells	Pretreatment of recipient	Primary response						Secondary response					
		day 8		15		22		day 6		11		16	
		AA	CRI_A	AA	CRI_A	AA	CRI_A	AA	CRI_A	AA	CRI_A	AA	CRI_A
Naive	None	–	–	–	–	–	–	9.8	–	27.0	–	19.1	–
Bone marrow	KLH-primed	0.5	–	1.5	–	1.9	–	139.7	–	466.0	–	387.2	–
Immune	None	0.6	–	3.3	5.6	3.2	5.6	32.5	16.4	81.4	37.4	81.2	37.0
Bone marrow	KLH-primed	3.4	1.0	36.5	43.0	99.7	29.0	185.6	53.4	667.8	146.3	554.7	108.7

AA, anti-arsonate antibodies (μg/ml); CRI_A, cross-reactive idiotype$_A$-positive antibodies (μg/ml); –, <0.2 μg/ml

[a] Bone marrow cells from normal or immunized A/J mice were transferred into irradiated syngeneic A/J mice primed or not with carrier (KLH). Similar results were obtained with spleen cells. The disappearance of CRI was noted in 32 out of 33 mice

There is nothing remarkable in the preimmune repertoire of A/J mice. By contrast, the anti-arsonate repertoire is much larger than the CRI repertoire. The CRI_A repertoire accounts for only 0.5% of the total arsonate repertoire. Nonetheless, after arsonate immunization, the CRI repertoire in A/J mice represents 50%–70% of the arsonate repertoire.

The frequencies of CRI_A or CRI_A-like precursors in both A/J and Balb/c strains are essentially the same. In fact, to our knowledge, the frequency of precursors in several systems of recurrent Id is not much different from the frequency of "private" or "silent" Id. On the other hand, all efforts to measure the frequency of B-cell precursors in A/J mice able to secrete auto-anti-Id antibodies have failed. This is true not only for the arsonate system but also for the Np system (K. Rajewsky, personal communication). Thus, Id dominance cannot be explained by gene amplification, by the frequency of precursors in the preimmune repertoire, or by a high affinity of CRI_A for arsonate.

There are also some new puzzling data from our laboratory.

When normal spleen cells or bone marrow cells from naive A/J mice are injected into syngeneic irradiated A/J mice together with Ars-KLH (Table 1), the response to arsonate is not abolished, but rather, the CRI_A Id disappears (even after several injections of the antigen). In this case, the gene is present, as are in the precursors able to secrete the CRI_A Id, but the dominance is lost. This observation rules out the simple hypothesis that the necessary and sufficient condition for the emergence of a recurrent Id is the presence of a "suitable" germ-line gene.

Let us now assume that the Id network is physiological (see below) with pairs of interacting Id, Ab1 and Ab2. Again, the terms Id and anti-Id are purely operational, and the immune system does not identify a particular immunoglobulin as one or the other; it knows only that some immunoglobulins recognize other immunoglobulins. In a general sense, Ab1 can be considered as a "brake" for AB2 and vice versa. If one of these interacting partners in an antiself antibody, it (e.g., Ab2) would be suppressed by its anti-Id antibody (e.g., Ab1, the idiotype!). There is no more reflection in the mirror, because the mirror has been blurred. We propose that recurrent Id are, in fact, anti-Id antibodies for antibodies directed against polymorphic self-antigens. (This has nothing to do with internal images.) The system is already "programmed to produce Ab1" (the recurrent Id), because Ab2 is suppressed. This is consistent with the fact that we cannot detect precursors able to secrete Ab2.

The other Id recognizing the same external antigen will be less favored, because their "preferred" partners are not suppressed.

We remind those who consider irradiated animals as "in vivo culture tubes" for donor cells that a major effect of irradiation is cell death, which, in turn, leads to the release of tremendous amounts of self-antigens.

4 The Arsonate Saga and the Learning of Self-Recognition

Another advantage of the study of the arsonate system is the existence of a suppressor cascade, Ts_1, Ts_2, Ts_3 (GREENE et al. 1982), which mirrors the Id cascade of the antibodies Ab1, Ab2, Ab3. Briefly, Ts_1 releases factors (TsF1) that are idiotypically crossreactive with CRI_A positive antibodies. These factors (or receptors?) are able to induce Ts2 cells which are anti-Id. Again, Ts2 cells can induce the effectors, Ts3, which are idiotypic and can bind antigen. It has been shown that neonatal anti-μ treatment can profoundly alter the suppressor cascade and particularly the Igh restriction (SY et al. 1981). We therefore investigated the induction of a silent Id which can induce alterations in the Id suppressor circuit. Balb/c mice were injected with rabbit anti-CRI_A in order to induce the expression of CRI_A-like Id. This treatment, in fact, enabled Balb/c mice to respond to A/J first-order suppressors. Untreated Balb/c mice are not susceptible to the suppressors of A/J. The same Id manipulation allows Balb/c mice to produce specific TsF1, displaying some crossreactive Id with CRI_A-positive antibodies. Thus, the induction of a silent Id has profound effects on the T-cell suppressor cascade (SLAOUI et al. 1985a).

Parallel Id preprogramming both in A/J and manipulated Balb/c mice simultaneously at the T- and B-cell levels raises questions about the functional linkage between T- and B-cell repertoire and about the nature of the driving force leading to such preprogramming. Since it is now clear that T- and B-lymphocyte receptors are the products of different V-region pools (SIU et al. 1984), it is unreasonable to expect that antigen alone can induce similar Id specificities at both the B- and T-suppressor levels. Therefore, all models invoking clonal selection mediated by antigen alone to explain immunoregulation are insufficient to account for these data. It seems, therefore, that T cells are not the leaders of the immunological orchestra and that B cells are not just "the slaves" of T cells. Analogous to the H-2 system, a learning process seems to be involved in Id recognition. Like H-2 restriction, Id recognition is clearly not rigidly dictated by the genotype (SLAOUI et al. 1985a).

5 Thoughts on the Suppressor and Idiotypic Cascade

We know that the immune system of an individual can become committed to producing a given Id after antigen injection and to producing a predetermined Id by the Id cascade. It appears that suppression is dominant in the immune system and that any immune response is an escape from suppression. Is it necessary to postulate the presence of thousands of suppressors, each specific for one Id, or can we develop a simpler model?

In this section, we propose such a model, the purpose of which is to link antigenic bridges and Id networks and to add a new perspective on the cascade of suppressors.

We reject the hypothesis that available repertoires are those which are germline encoded. While that model may in part be accurate, it is most likely oversimplified. Consider instead the emergence of one Id (the problem of the initial choice is discussed in the final section), the expression of which will promote the induction of a suppressor cascade such as Ts1, Ts2, Ts3. The inducers and the effectors bear cross-reactive Id receptors and they can bind antigen. So they can make antigenic bridges. The transducer or Ts2 is anti-Id. The effector Ts3, which binds antigen, can in fact suppress *all* Id directed against the same antigen. Memory B cells are much less susceptible to suppression than naive B cells, so that suppression will be much more efficient for Id not chosen than for the one expressed. We do not need lots and lots of Id-specific suppressor T cells. Ts3 is sufficient. Thus, the choice of one Id allows the suppression of all the other Id. The remarkable fidelity of the immune system to the past, i.e., original antigenic sin and original Id sin, becomes easily interpretable with this model without myriad suppressor T cells specific for all the other Id. A detailed discussion of this model will be presented elsewhere.

In summary, a model based on the assumption that idiotypic adaptive differentiation exists, drastically changes the current picture of the immune system. Interactive dynamics becomes the rule. Each lymphocyte communicates with other lymphocytes and is thereafter permanently changed. In this respect, immunology begins to resemble embryology, neurobiology, and some parts of modern physics. Perhaps, a new language designed to fashion complexity is urgently needed (Thomas 1977).

Thus, from a simple operational definition of Id, based on its reactivity with an anti-Id serum, we have progressed to a view in which Ab1 and Ab2 can be called Ab2 and Ab1, where not all Id are equal, and where some idiotopes are more regulatory than others (Bona and Paul 1981), although the regulatory function of one idiotope can appear, disappear, reappear, or die, depending on the dynamics of the immune system. At that point, classical definitions become outmoded and, at best, confusing. In this respect, the epibody story is quite illuminating (Bona et al. 1984; Bona 1985).

6 On the Way to Idiotypic Vaccines

The induction of predominantly Ab1-like Id upon injection of one Ab2 clearly indicates that Ab2 can replace antigen. Thus, a new approach to vaccination emerges, which exploits elements of the immune system itself (BINZ and WIGZELL 1977; TILKIN et al. 1981), rather than the use of killed or inactivated viruses and bacteria to induce protective immunity. In this effort, recent results in different systems such as *E. Coli,* reovirus, poliovirus, Hepatitis B, rabies viruses, and tumors (ERTL and FINBERG 1984; STEIN and SÖDERSTRÖM 1984; SHARPE et al. 1984; UYTDEHAAG and OSTERHAUS 1985; NEPOM et al. 1984; REGAN et al. 1983; DE FREITAS et al., this volume) have been quite encouraging. Most of these studies have involved a search for protective recurrent Id. We suggest in this section that the same approach is feasible with Ab2 directed against "private" Id. Portions of this work have recently been published (FRANCOTTE and URBAIN 1984).

We raised rabbit anti-Id antibodies directed against nonrecurrent Ab1 anti-tobacco mosaic virus (TMV) antibodies from rabbits. These conventional Ab2 alone or coupled with LPS were injected into Balb/c mice, which do not express the rabbit Id after TMV injection. As a result, the mice treated with rabbit Ab2 synthesize the operationally defined Ab3, a large proportion of which binds TMV. Thus, these mice, which have never been exposed to TMV, synthesize anti-TMV antibodies after injection of a conventional Ab2 that does not behave at all like an internal image. Apparently, the Id cascade can cross the species barrier.

In those experiments, the polyclonal activator is targeted to lymphocytes displaying Ab1-like receptors. Interestingly, these mouse anti-TMV antibodies have been isolated and injected into rabbits or mice to obtain anti-Id antibodies. Operationally, we are at the fourth level of antibodies (Ab4). These Ab4 recognized specifically the initial rabbit Id (anti-TMV), but not the anti-TMV antibodies synthesized by other rabbits.

All the preceding manipulations used nonphysiological conditions (adjuvant, polyclonal activation, large amounts of Ab2).

We then repeated these experiments using low amounts of Ab2 (<1 μg) and without the use of nonphysiological adjuvants. We used the same rabbit anti-Id antibodies as in the Ab2-LPS experiments. We isolated dendritic cells or macrophages from naive Balb/c mice, pulsed these cells in vitro with rabbit Ab2, and injected them back into naive mice of the same strain. The mice that were injected with pulsed Ab2 dendritic cells and that had never seen the virus (TMV) synthesized Id-positive anti-TMV antibodies (results not shown). Pulsed macrophages were much less efficient. Subsequently all these mice received TMV for the first time. Mice which received pulsed dendritic cells synthesized high amounts of anti-TMV antibodies (the concentration was as high as 300 μg/ml).

As shown in Table 2, a normal primary response to TMV is in the range of 10 μg/ml. Injection of 0.5 μg Ab2 alone leads to the induction of Id-positive antibodies, but no enhancement is seen after TMV injection.

The data obtained with animal viruses such as reoviruses, rabies, and Sendai viruses indicate that a slightly larger dose (1.25–2.5 μg) of anti-Id antibody injected into mice in the presence of adjuvants induces formation of not only virus-binding antibodies but also antibodies protective against viral challenge.

Table 2. Concentration of Id-positive antibodies and anti-TMV antibodies (μg/ml) in mice injected with pulsed dendritic cells or with pulsed macrophages followed by antigen immunization

Treatment of mice before TMV injection	Concentration of antibody (μg/ml) in immunized mice	
	Id^+	Anti-TMV
No. of Ab2-pulsed DC		
10^5	290±170	274±64
5×10^4	510±147	305±103
2×10^4	275±110	203±53
10^4	75±20	68±24
No. of Ab2-pulsed Mø		
10^5	60±13	31±13
5×10^4	46±18	27±11
2×10^4	35±6	8±5
Amount of Ab2 (μg)		
0.5	10±2	8±2
2 Ab2	8±4	7±3
None	<0.5	8±3

DC, dendritic cells; Mø, macrophages

7 Is the Immune System a Functional Idiotype Network?

Every immunologist, even the most conservative, is convinced that an immune response is finely tuned by a series of complex cell interactions. The controversial issue is whether clonal selection of interacting lymphocytes is due only to antigens or also to Id interactions that occur even in the absence of exogenously injected Id or anti-Id.

Let us take the case of the Id cascade as an example. Some argue that these results have nothing to do with a functional Id network. The Id cascade process demonstrates the power of selection (a double selection with anti-Id antibodies and with antigen). This process operates on the "inherent fitness" of germ-line receptors (affinity for antibodies) and on the adaptibility of these germ-line encoded structures (the ability to generate "useful" somatic variants). If this were the case, our experiments would reflect the thinking of COHN, who rejects the notion of an Id network. Did we accomplish the reverse of what we wanted to accomplish? Even if this were true, it was a useful approach, since we learned how to manipulate immune responses in a specific way. We are not very interested in nonspecific enhancement or suppression, which would be like approaching archeology with bulldozers.

The power of these manipulations rests in the fact that a major part of Ab3 is Ab1-like and part of Ab4 behaves like Ab2. This strange mathematics

(1 = 3; 4 = 2) is extremely difficult to understand without postulating the existence of a functional Id network. Can we now prove that Id selection is operative physiologically? Several sets of experiments shed light on this question. First, EICHMANN and RAJEWSKY (EICHMANN and RAJEWSKY 1985) have shown that minute amounts of a perturbing agent (physiological amounts) can strongly modify the outcome of an immune response. Second, the multihit curves of EICHMANN et al. (1983) cannot be understood without assuming specific interactions deep inside the system. Then there is the effect of maternal immunoglobulin Id (Ab3 immunoglobulins). In this case, the mothers are manipulated before being mated with naive males. The offspring acquire the "learning" of the mothers by physiological means (transfer by the placenta; WIKLER et al. 1980).

The modification of available repertoires at the B-cell level also leads to dramatic changes in functional repertoires of T cells, despite the fact that immunoglobulin receptors and T-cell receptors are encoded by distinct multigene families.

Furthermore, as explained above, the CRI_A dominance is lost in syngeneic transfer experiments. In this case, no Id or anti-Id Id was injected and the corresponding idiotypically related suppressor cascade has been modified (SLAOUI et al., to be published). All of these results suggest that Id selection is performed by the immune system itself and not by the research worker. If this is correct, one can say that the immune system is a functional Id network, incorporating selection, based on the inherent fitness or adaptibility, which is due not only to the ability to sustain the generation of "useful" somatic variants, but also to the permissiveness of the regulatory circuits. This permissiveness depends on the previous history of the system, a major property of networks.

8 Order from the Beginning or Order out of Chaos?

In the first part, we discussed the notion of germ-line circles: the idea that Id networks are already encoded in the germ-line. This is what we call order from the beginning.

At first sight, this idea does not fit with DNA polymorphism of immune multigene families. We know that all multigene families are unstable and are the constant target of recombination, gene conversion (or gene talking). But all that is needed is the conservation of complementary families of immunoglobulins. After all, the genes coding for the α and β chains evolve constantly by the association of α and β chains is maintained, as well as the binding to heme. Furthermore, all these processes of gene recombination and gene conversion tend to homogenize multigene families and thus to reduce diversity. The germ-line encoded network could be the internal selective pressure which guarantees the maintenance of diversity.

Another point of view can be offered. There is no germ-line encoded Id network. The network is just a consequence of diversity. Given the tremendous potential repertoire of the immune system, it is difficult to escape the idea that in such a chaos of diversity, the immune system cannot avoid the recognition of itself. It is not too difficult a task to imagine models in which order will emerge from this chaos. The preimmune repertoire will be confronted with thousands of self-antigens and maternal Id will shape some elements of the network.

Autoregulation will emerge from Id adaptive differentiation. We have suggested a model for the emergence of recurrent Id. We are still left with the question: What about idiotypes à la Oudin? Why do different rabbits make different Id when confronted with the same antigen? After all, Id à la Oudin are much more numerous than recurrent Id, even if the relevant literature gives the contrary impression.

Some years ago, Oudin told us that in science, facts must be reproducible. Idiotypy, by essence is irreproducible, unless we use the idiotypic cascade. In fact, however, J. Oudin decided to publish his famous 1969 paper because he could reproduce the irreproducible.

What determines the choice of one Id? Is this choice "preprogrammed" (like for recurrent Id)? Is it possible to solve this problem by experimental procedures?

For a long time ago now, we have been dreaming about an impossible experiment. We inject a given rabbit with a given antigen, let us say in November. We raise an Ab2 which identifies the Id. What would have been the result if the rabbit was injected with the same antigen in the same rabbit several months earlier or several months later? The same Id or different Id?

This experiment, as such, is impossible to do because we know that once the choice has been made, the choice is stable. All we know is that the "choice" can be imposed by maternal Ab3 antibodies or by injecting an Ab2 chosen by the research worker. Possibly, the effects of maternal Ab3 can be explained by the suppression of an Ab2. The "blind spots" in the Ab2 repertoire will favor the emergence of complementary Ab1. Once again, the mirror has been blurred. We could say that self-antigens are different between different individuals from an outbred species. It is conceivable that different individuals with different self-antigens, different maternal immunoglobulins, and different antigenic histories will synthesize different Id.

This explanation is not really satisfactory because the most recent estimates of immunoglobulin diversity (combining combinational joining of germ-line genetic segments, fluctuations at the border of recombination, insertion of N segments, somatic mutations, gene "talking"...), suggest that the number of potential antibodies is possibly greater than the total number of lymphocytes. If this is correct, it is hard to avoid the idea, that the "idiotypic choice" could be partly stochastic. So we are lost somewhere in the twilight zone between chaos and order.

After all, we have learned (or we have to learn) modesty. Deep inside, we know that we shall never completely understand the complexity of networks. This has already been beautifully stated by Jerne: "Those who always seek exterior pressures (e.g., microbes) to account for the evolution of the sets of V genes would do well to turn their vision towards the interior of themselves and there discover the mystery, perhaps never completely revealable, of the immune system."

Our search or research has been, is, and will forever be one "unended quest." We are convinced that confusion is much more in our minds than in the immune system itself.

Order from the beginning or order out of chaos? It is not impossible to combine the two. The germ-line would encode a network of low connectivity, and the chaos could increase this basic low connectivity and improve the adaptability of the system.

Acknowledgements. This paper, which was written too rapidly, is dedicated to Henry Kunkel who died too early, to Niels Kaj Jerne – the greatest immunologist of all time – and to Jacques Oudin. All three gave me much advice during the last years. It is a special pleasure to acknowledge Don Capra, Mark Greene, Bary Benacerraf, Constantin Bona, Pierre-André Cazenave, Antonio Coutinho, Michel Fougereau, Fritz Melchers, Gino Doria, Heinz Köhler, and C. Milstein for cooperation, help, advice, and discussions. I cannot adequately express my appreciation to R. Jeener, J. Brachet, R. Thomas, H. Chantrenne and I. Prigogine. I was and remain their student. I thank my collaborators for their unending help. I shall always remember long discussions with H. Wigzell and R. Gershon. I should not forget my secretary, Mrs. Neirinckx, for her constant help. I am very much indebted to the Basel Institute of Immunology, which has several times provided me an exciting place for discussion. Thanks also to Lee Hood and Melvin Cohn for expert criticism. Lastly, I thank the Actions de Recherches Concertées from the Belgian State, the F.N.R.S., la Fondation Van Buuren, and la Caisse Générale d'Epargne et de Retraite for funding this research.

References

Binz H, Wigzell H (1977) Antigen-binding, idiotypic T lymphocytes receptors. Contemp Top Immunobiol 7:113–177

Bona C (1984) Regulatory idiotypes. In: Köhler G, Urbain J, Cazenave PA (eds) Idiotypy in biology and medicine. Academic, New York, pp 29–41

Bona C (1985) Epibodies: a particular set of antiidiotypes specific for antiantibodies. Clin Immunol (to be published)

Bona C, Heber-Katz H, Paul WE (1981) Idiotype-antiidiotype regulation. I. Immunization with a levan-binding myeloma protein leads to the appearance of autoanti-(anti-Id) antibodies and to activation of silent clones. J Exp Med 153:951–967

Bona C, Goldberg B, Metzger D, Urbain J, Kunkel H (1984) Anti-immunoglobulin antibodies. IV. Cross-reaction of anti-idiotypic antibodies specific for rabbit and murine anti-A_1 allotype antibodies with Fc fragment of human immunoglobulins. Eur J Immunol 14:548–552

Cancro M, Thompson MA, Raychaudhuri S, Hilbert D (1984) Ontogeny of the HA-responsive B-cell repertoire: interaction of heritable and inducible mechanisms in the establishment of phenotype. In: Köhler G, Urbain J, Cazenave PA (eds) Idiotypy in biology and medicine. Academic, New York, pp 144–170

Capra JD, Fougereau M (1984) Structural correlates of idiotypes. Ann Immunol (Paris) 135C:7

Capra JD, Slaughter CA, Milner ECB, Estess P, Tucker PT (1982) The cross-reactive idiotype of A-strain mice. Serological and structural analyses. Immunol Today 3:332

Cazenave PA (1977) Idiotypic-antiidiotypic regulation of antibody synthesis in rabbits. Proc Natl Acad Sci USA 74:5122–5125

Cohn M (1981) Conversation with Niels Kaj Jerne. Cell Immunol 61:425–436

Eichmann K, Rajewsky K (1985) Induction of T and B cell immunity by antiidiotypic antibody. Eur J Immunol 5, 661–667

Eichmann K, Fey K, Kuppers R, Melchers I, Simon MM, Weltzien HV (1983) Network regulation among T cells; conclusions from limiting dilution experiments. Springer Seminars in Immunopathology 6, 7–32

Ertl HC, Finberg RW (1984) Sendai virus spcific T-cell clones: induction of cytolitic T-cells by an anti-idiotypic antibody directed against a helper T cell clone. Proc Natl Acad Sci USA 81:2850–2854

Francotte M, Urbain J (1984) Induction of anti-tobacco mosaic virus antibodies in mice by rabbit antiidiotypic antibodies. J Exp Med 160:1485–1494

Greene M, Nelles A, Sy MS, Nisonoff A (1982) Regulation of immunity to the azobenzenearsonate hapten. Adv Immunol 32:243

Jerne NK (1972) What precedes clonal selection in ontogeny of acquired immunity? Ciba Symposium. Elsevier, Amsterdam, pp 1–15

Jerne NK (1974) Towards a network theory of the immune system. Ann Immunol (Paris) 125C:373–389

Jerne NK (1984) Idiotypic networks and other preconceived ideas. Immunol Rev 79:1–29

Jerne NK, Roland J, Cazenave PA (1982) Recurrent idiotopes and internal images. EMBO J 1:243–247

Kindt TJ, Seide RK, Bokish VA, Krause RM (1973) Detection of idiotypic crossreaction among streptococcal antisera from related rabbits. J Exp Med 138:522–537

Kuettner MG, Wang A, Nisonoff A (1972) Quantitative investigations of idiotypic antibodies. VI. Idiotypic specificities as a potential genetic marker for the variable region of mouse immunoglobulin polypeptide chain. J Exp Med 135:579–595

Kunkel HG, Mannick M, William RC (1963) Individual antigenic specificities of isolated antibodies. Science 140:1218–1219

Leo O, Slaoui M, Marvel J, Milner E, Hiernaux J, Moser M, Capra D, Urbain J (1985) Idiotypic analysis of polyclonal and monoclonal anti-p-azophenylarsonate antibodies of Balb/c mice expressing the major crossreactive idiotype of the A/J strain. J Immunol 134:3

Lucas A, Henry C (1982) Expression of the major crossreactive idiotype in a primary anti-azobenzenearsonate response. J Immunol 128:802

Manser T, Huang S-Y, Gefter M (1985) The influence of clonal selection on the expression of immunoglobulin variable region genes. Science (to be published)

Meek K, Jeske D, Slaoui M, Leo O, Urbain J, Capra D (1984) Complete amino acid sequence of heavy chain variable regions derived from two monoclonal anti-p-azophenylarsonate antibodies of Balb/c mice expressing the major crossreactive idiotype of the A/J strain. J Exp Med 160:1070–1086

Moser M, Leo O, Hiernaux J, Urbain J (1983) Idiotypic manipulation in mice: Balb/c mice can express the crossreactive idiotype of A/J mice. Proc Natl Acad Sci USA 80:4474–4479

Near R, Manser T, Gefter M (1985) The generation of major and minor idiotype-bearing families of anti-p-azophenylarsonate antibodies: the stochastic utilization of VH gene segments. J Immunol (to be published)

Nepom O, Nelson K. Holbeck S, Hellström I, Hellström K (1984) Induction of immunity to a human tumor marker by in vivo administration of antiidiotypic antibodies in mice. Proc Natl Acad Sci USA 81:2864

Oudin J, Bordenave G (1971) Idiotypy of rabbit antibodies against salmonella abortus-equi. Nature 231:86

Oudin J, Cazenave PA (1971) Similar idiotypic specificities in immunoglobulin fractions with different antibody functions or even without detectable antibody function. Proc Natl Acad Sci USA 68:2616–2620

Oudin J, Michel M (1963) Une nouvelle forme d'allotypie des immunoglobulines du sérum de lapin. C.R. Seances Acad Sci 257:805–808

Oudin J, Michel M (1969) Idiotypy of rabbit antibodies. J Exp Med 130:595–1006

Pollock B, Kearney JF (1984) Identification and characterization of an apparent germline set of auto-antiidiotypic regulatory B lymphocytes. J Immunol 132:114–119

Regan KJ, Wunner WH, Wiktor T, Koprowski H (1983) Antiidiotypic antibodies induce neutralizing antibodies to rabies virus glyco proteins. J Virol 48:660–666

Sachs DL, Esser K, Sher A (1982) Immunization of mice against african trypanosomiasis using antiidiotypic antibodies. J Exp Med 155:1108–1119

Schnurr I (1981) Idiotypes. Antigens on the inside. Workshop at the Basel Institute for Immunology, edited by Hoffmann La Roche, Basel, nov. 19–20. pp 122–130

Sharpe A, Gaulton G, Mc Dade K, Fields B, Greene M (1984) Syngeneic monoclonal antiidiotype can induce cellular immunity to reovirus. J Exp Med 160:1195–1205

Siekewitz M, Geter ML, Brodeur P, Riblet R, Mashak-Rothstein A (1982) The genetic basis of antibody production: the dominant antiarsonate idiotype response of the strain-A mouse. Eur J Immunol 12:1023

Sigal NH (1982) Regulation of azophenylarsonate specific repertoire expression. J Exp Med 156:1352

Siu G, Kronenberg M, Strauss E, Haars R, Mak T, Hood L (1984) The structure rearrangement and expression of dβ gene segments of the murine T-cell antigen receptor. Nature 311:344

Slaoui M, Leo O, Marvel J, Moser M, Hiernaux J, Urbain J (1984) Idiotypic analysis of potential and available repertoires in the arsonate system. J Exp Med 160:1–11

Slaoui M, Urbain J, Lowy A, Monroe J, Benacerraf B, Greene M (1985a) Antiidiotypic treatment of Balb/c mice induces CRI_A bearing suppressor cells and changes the predicted Igh restriction pattern (to be published)

Stein K, Söderström J (1984) Neonatal administration of idiotype or antiidiotype primes for protection against Escherichia coli K13 infection in mice. J Exp Med 160:1001–1011

Sy MS, Nisonoff A, Germain R, Benacerraf B, Greene M (1981) Antigen and receptor driven regulatory mechanisms. VIII. Suppression of idiotype negative *p*-azobenzenearsonate specific T cells results from the interaction of an antiidiotypic second order T suppressor cell with a cross-reactive idiotype positive *p*-azobenzenearsonate primed T cell target. J Exp Med 153:1415

Thomas R (1977) Kinetic logic. A boolean approach to the analysis of complex regulatory systems. Lecture notes in biomathematics, vol 29. Springer, Berlin Heidelberg New York

Tilkin AF, Schaaf-Lafontaine N, Van Acker A, Boccadoro M, Urbain J (1981) Reduced tumor growth after low-dose irradiation or immunization against blastic suppressor T cells. Proc Natl Acad Sci USA 78, 3:1809–1812

Urbain J (1974) Cellular recognition and evolution. Arch Biol 85:139–150

Urbain J (1976) Idiotypes expression of antibody diversity and network concepts. Ann Immunol (Paris) 127C:357–374

Urbain J, Van Acker A, De Vos-Cloetens C, Urbain-Vansanten G (1972) Increase and decrease in binding affinity of antibodies during the immune response. Immunochemistry 9:121–136

Urbain J, Wikler M, Franssen J-D, Collignon C (1977) Idiotypic regulation of the immune system by the induction of antibodies against antiidiotypic antibodies. Proc Natl Acad Sci USA 74:5126–5130

Urbain-Vansanten G, Van Acker A, Tasiaux N, De Vos-Cloetens C, Urbain J (1979) Synthesis of antibodies and immunoglobulins learning recipient allotypic markers and donor idiotypic specificities in irradiated rabbits grafted with allogeneic cells from hyperimmune donors. Ann Immunol (Paris) 130C:397–406

Uytde Haag F, Osterhaus DM (1985) Induction of neutralizing antibody in mice against poliovirus type II with monoclonal antiidiotypic antibodies. J Immunol (to be published)

Wikler M, Urbain J (1984) Idiotypic manipulation of the rabbit immune response against micrococcus luteus. In: Köhler, Urbain J, Cazenave PA (eds) Idiotypy in biology and medicine. Academic, New York, pp 219–241

Wikler M, Franssen J-D, Collignon C, Leo O, Mariamé B, Van de Walle P, Degroote D, Urbain J (1979) Idiotypic regulation of the immune system. J Exp Med 150:184–195

Wikler M, Demeur C, Dewasme G, Urbain J (1980) Immunoregulatory role of maternal idiotypes. J Exp Med 152:1024–1035

Wysocki L, Margolies M, Huang B, Nemaze D, Wechsler D, Sato V, Smith J, Gefter M (1985) Combination diversity within variable regions bearing the predominant anti-arsonate idiotype of A strain mice. J Immunol (to be published)

MIX
Papier aus verantwortungsvollen Quellen
Paper from responsible sources
FSC® C105338

If you have any concerns about our products,
you can contact us on
ProductSafety@springernature.com

In case Publisher is established outside the EU,
the EU authorized representative is:
Springer Nature Customer Service Center GmbH
Europaplatz 3, 69115 Heidelberg, Germany

Printed by Libri Plureos GmbH
in Hamburg, Germany